T0271876

Containerization in Maritime Transport

Containerization provides optimization of handling processes in terms of intermodality and efficient cargo handling, and for maritime transport, in particular, it provides further optimization of shipping processes in terms of volume and distance. Containerization has become the most significant factor stimulating the development of modern global trade. With the progress of globalization taken into account (longer distances and increasing cargo volumes), it can be seen that cargo becomes predestined to be transported by sea, which encourages shipowners to enlarge their fleets of container ships. *Containerization in Maritime Transport: Contemporary Trends and Challenges* addresses the key challenges to maritime transport and containerization, beginning with economic and managerial factors through organizational, technical, operational, information and IT challenges and ending with ecological challenges—ideally to lessen the environmental impacts of maritime transport.

Features:

- Discusses the latest technological advances in shipping, including augmented reality (AR), virtual reality (VR), artificial intelligence (AI), 5G networks, smart camera and computer vision systems, and digital twin technology.
- Presents ecological considerations and solutions that are indispensable to develop efficient and safe green supply chains.
- Examines the economic aspects of shipping with regard to transport and container handling costs in international trade.

Containerization in Maritime Transport

Contemporary Trends and Challenges

Ryszard Miler, Eugeniusz Gostomski,
and Tomasz Nowosielski

CRC Press
Taylor & Francis Group
Boca Raton London New York

CRC Press is an imprint of the
Taylor & Francis Group, an **informa** business

Designed cover image: Shutterstock | Travel mania

First edition published 2023
by CRC Press
6000 Broken Sound Parkway NW, Suite 300, Boca Raton, FL 33487–2742

and by CRC Press
4 Park Square, Milton Park, Abingdon, Oxon, OX14 4RN

CRC Press is an imprint of Taylor & Francis Group, LLC

© 2023 Taylor & Francis Group, LLC

Translated by Małgorzata Wilska

Library of Congress Cataloging-in-Publication Data
Names: Miler, Ryszard K., author. | Gostomski, Eugeniusz, author. | Nowosielski, Tomasz, author.
Title: Containerization in maritime transport : contemporary trends and challenges / Ryszard Miler, Eugeniusz Gostomski, and Tomasz Nowosielski.
Description: Boca Raton : CRC Press, 2023. | Includes bibliographical references and index.
Identifiers: LCCN 2022028956 (print) | LCCN 2022028957 (ebook) | ISBN 9781032360713 (hbk) | ISBN 9781032360720 (pbk) | ISBN 9781003330127 (ebk)
Subjects: LCSH: Containerization. | Shipping.
Classification: LCC TA1215 .M55 2023 (print) | LCC TA1215 (ebook) | DDC 387.5/442—dc23/eng/20221006
LC record available at https://lccn.loc.gov/2022028956
LC ebook record available at https://lccn.loc.gov/2022028957

ISBN: 978-1-032-36071-3 (hbk)
ISBN: 978-1-032-36072-0 (pbk)
ISBN: 978-1-003-33012-7 (ebk)

DOI: 10.1201/9781003330127

Typeset in Times
by Apex CoVantage, LLC

Contents

PART 1 Maritime Containerized Transport (the Status Quo of Its Structure and Main Processes)

PART 2 Economic Challenges to Maritime Containerized Transport

PART 3 Managerial and Operational Challenges to Maritime Containerized Transport

PART 4 *Sustainable Development Challenges to Containerized Maritime Transport*

About the Authors

 **UG Professor Emeritus Eugeniusz Gostomski**, Ph.D., is a long-standing academic employee of the Department of International Business at the Faculty of Economics at the University of Gdańsk and a member of a supervisory board of a stock exchange-listed company. He specializes in the field of economic analysis and international maritime trade, with particular consideration on the German economy. He is the author of numerous monographs, including *German Economy*, *International Banking*, and *International Maritime Trade*.

 **WSB Associate Professor Ryszard Miler**, Ph.D., is a director of the Baltic Sea Centre of Applied Logistics at the Faculty of IT and New Technologies at WSB in Gdańsk and an academic employee of the WSB-DSW Scientific Federation. He holds the title of European Senior Logistician, awarded by the European Logistics Association. He specializes in the field of logistics management, telematics systems and maritime transport. He is an author of numerous scientific monographs, including *Safety and Security of Maritime Transport* and *Telematics Solutions in Maritime and Inland Waterway Transport*.

 **Assistant Professor Tomasz Nowosielski**, Ph.D., is an academic employee of the Department of Transport and Maritime Trade at the Faculty of Economics at the University of Gdańsk. His scientific interests include land and maritime transport systems, globalization processes and the protection of the marine environment. He is a co-author of several monographs, including *International Maritime Trade* and *Evolution and Significance of Maritime Ports in EU Countries*.

Acknowledgments

The authors wish to thank the Authorities of the University of Gdańsk (Poland) and the Research Federation of WSB & DSW Universities (Poland) for their organizational and financial support for this monograph.

The authors wish to thank to Michał Janczewski for his great support full of professionalism in preparation of all graphics for our monograph.

Glossary of Terms and Abbreviations (Acronyms)

AIS	Automated Identification System
ALARP	As Low as Reasonably Practicable
AMOS	Asset Management Operating System
AMS	Advanced Manifest Surcharge
ARPA	Automatic Radar Plotting System
ATA	Actual Time of Arrival
ATD	Actual Time of Departure
AtoN	Aid to Navigation
BAF	Bunker Adjustment Factor
BST	Basic Safety Training
C/S lub PCS	Congestion Surcharge
CAF	Currency Adjustment Factor
CBM	Condition-Based Monitoring
CBR	Commodity Box Rate
CFS	Container Freight Station
EDI	Electronic Data Interchange
CG	Contracting Government
CG	Coast Guard
CISE	Common Information Sharing Environment
CleanSN (CSN)	CleanSeaNet
COBO	Coal-Ore-Bulk-Oil
CoS	Certificates of Safety
CRS	Coastal Radar Surveillance
CSN	CleanSeaNet
CSO	Company Security Officer
DGLM	Dangerous Goods Location and Maintenance
DGMT	Dangerous Goods Maritime Traffic
DGPS	Differential Global Positioning System
DGRA	Dangerous Goods Risk Analysis
DGSC	Dangerous Goods Shipping Conditions
DOS	Declaration of Security
ECDIS	Electronic Chart Display and Information System
ECTS	Electronic Container Tracking Service
EDI	Electronic Data Interchange
EEDI	Energy Efficiency Design Index
EEOI	Energy Efficiency Operational Indicator
EEZ	Exclusive Economic Zone
EGNOS	European Geostationary Navigation Overlay Service
EMCIP	European Marine Casualty Information Platform

EMSA	European Maritime Safety Agency
ENC	Electronic Navigational Chart
ERP	Enterprise Resources Planning
ERS	Equipment Repositioning Surcharge
EIS	Equipment Imbalance Surcharge
ERS	Electronic Reporting System
ETA	Estimated Time of Arrival
ETD	Estimated Time of Departure
FCL	Full Container Load
FoN	Freedom of Navigation
FSC	Flag State Control
GDP	Gross Domestic Product
GHG	Greenhouse Gas
GIS	Geographic Information System
GISIS	Global Integrated Shipping Information System
GL	Gear Logbook
GLONASS	Global Navigation Satellite System
GMDSS	Global Maritime Distress and Safety System
GPS	Global Positioning System
GT	Gross Tonnage
HAZMAT	Ships Hazardous Materials
HELCOM	Helsinki Commission
HOA	Horn of Africa
HVG	High Value Goods
IACS	International Association of Classification Societies
IALA	International Association of Lighthouse Authorities
IBS	Integrated Bridge System
ICF	International Compensation Fund for GHG emissions from Ships
IEM	Information Exchange Model
IMDatE	Integrated Maritime Data Environment
IMDG Code	International Maritime Dangerous Goods Code
IMO	International Maritime Organization
INS	Integrated Navigation System
ISM (Kodeks)	International Safety Management Code
ISO	International Organization for Standardization
ISP	Integrated Ship Profile
ISPS Code	International Ship and Port Facility Security Code
IT/ICT	Information Technology/Information and Communication Technology
IWRAP	IALA Waterway Risk Assessment Programme
KSBM	Krajowy System Bezpieczeństwa Morskiego, National System of Maritime Safety and Security (Poland)
LASH	Lighter Aboard Ship Vessels
LCL	Less Container Load

LEO	Low Earth Orbital
LNG	Liquefied Natural Gas
LRFPR	Lloyd's Register—Fairplay Research
LRIT	Long Range Identification and Tacking
MARPOL 73/78	International Convention for the Prevention Pollution from Ships
MAS	Maritime Assistance Services
MEPC	IMO Marine Environment Protection Committee
MRV of CO$_2$	Monitoring, Reporting and Verification System
MSSIS	Maritime Safety and Security Information System
NAVTEX	NAVigational TEXt Messages
NPOC	Next Port of Call
NSA	National Shipping Authority
NSW	National Single Window
OECD	Organization for Economic Co-operation and Development
Paris MOU	Paris Memorandum of Understanding
PCMS	Propulsion Condition Monitoring Service
PFSA	Port Facility Security Assessment
PFSO	Port Facility Security Officer(s)
PFSP	Port Facility Security Plan
PSC	Port State Control
PSSA	Particularly Sensitive Sea Area
RFID	Radio Frequency Identification
RSO	Recognized Security Organization
RTW	Round the World
SAR	Search and Rescue
SCS	Suez Canal Surcharge
SECA	Sulphur Emission Control Area—SOx
SEEMP	Ship Energy Efficiency Management Plan
SLE	Sea Law Enforcement
SMS	Safety Management System
SOLAS 1974	Convention for the Safety of Life at Sea
SRES	Special Report on Emissions Scenarios (IPCC)
SRP	Ship Risk Profile
SSC	Security (Sur)charge
SSN	SafeSeaNet
SSO	Ship Security Officer
SSP	Ship Security Plan
SSS	Short Sea Shipping
STIRES	SafeSeaNet Traffic Information Relay and Exchange System
MRN	Movement Reference Number
OCR	Optical Character Recognition
OOG	Out of Gauge (Cargo)
POD	Port of Destination
PTI	Pre-Trip Inspection
RCMS	Robotic Container Management System

RMG	Rail Mounted Gantry (Crane)
RS	Reach Stacker
RTG	Rubber Tired Gantry (Crane)
SC	Straddle/Shuttle Carrier
STS	Ship to Shore (Crane)
SWL	Safe Working Load
TEN-T	Trans-European Transport Networks (corridors)
TOS	Terminal Operation System
TT/UTR	Terminal Tractor/Utility Tractor Rig
TEU	Twenty Equivalent Unit
THC	Terminal handling Charge
TTW	Territorial Waters
UNCLOS	United Nations Convention on the Law of the Sea
UNCTAD	United Nations Conference on Trade and Development
UNEP	United Nations Environment Programme
UNFCCC	United Nations Framework Convention on Climate Change
USD	User Specific Data
VDS	Vessel Detection System
VMS	Vessel Monitoring System
VTMIS	Vessel Traffic Monitoring and Information System
VTS	Vessel Traffic Service
WECDIS	Warship Electronic Chart Display and Information System
WIC	Worker Identification Card
WRP	War Risk Premium

Introduction

There are only two main factors in the holistic analysis of elements pertaining to modern transport in its holistic understanding that affect the potential of international (and more and more global) trade: containerization (which provides optimization of handling processes in terms of intermodality and efficient cargo handling) and maritime transport (which provides optimization of shipping processes in terms of volume and distance).

Considering the intensity of systemic changes in the field of transport, the appearance of containers is nowadays referred to as the third transport revolution (Neider and Marciniak-Neider 1997, 24) because containerization has become the most significant factor stimulating the development of modern global trade. The dynamic development of trade in the world generates additional demand for containerized cargo shipping. Taking the progress of globalization into account (longer distances and growing cargo volumes), cargo becomes predestined to be transported by sea and this fact, in turn, encourages shipowners to enlarge their fleets of container ships. The increased supply of maritime transport services (sailings, container slots) results in the secondary pressure imposed on trade. In this way, it is possible to observe a "self-winding" spiral of development (a sustainable path of growth) in the global trade volume, which partially explains the economic phenomenon of modern containerization.

In search of arguments for a more insightful explanation of the discussed phenomenon, it should be noted that in the field of technology and operationalization of processes, the use of intermodal containers has radically changed maritime transport of general cargo and has significantly affected the containerization of shipment in inland waterway transport—a complementary branch of water transport. The process of containerization has already resulted in the shortening of time when (container) vessels stay at ports and in the better use of vessel cargo loading space and the surface of port yards (terminals) because it is possible to store containers in multilayer stacks. The implemented shipping innovations that minimize the risk of cargo damage during shipping and cargo handling processes have also resulted in a decrease in cargo handling costs and have immensely increased the operational efficiency of processes.

However, the key factors (determinants) of the further development of containerization in maritime transport are the dynamic market environment (with the consideration of stochastic risk factors, such as the Covid-19 pandemic), advancing internationalization and standardization of trade relations (based on such standards as Incoterms or ISO), the globalization of international trade (measured by the growth of the foreign trade volume), the specific role of maritime transport in international trade (accounting for 90% of the volume of transported cargo), universality of information technology (IT) and information and communication technology (ICT) in telematics systems and pressure to protect the natural environment.

DOI: 10.1201/9781003330127-1

Maritime container transport has been changing under the influence of IT/ICT, and it is now possible to observe a decrease in the human-factor failure in the management and control of various processes. It has been achieved by the use of augmented reality (AR), virtual reality (VR), artificial intelligence (AI) and 5G networks, which allow users to apply smart camera systems and computer vision systems to provide full automation and to use digital twin technology, which is applied in the 3D visualization of processes in real time in the operation of facilities included in container terminal suprastructure.

Furthermore, considering modern global circumstances, it is impossible to maintain the proper level of competitiveness and capabilities pertaining to the development of containerization (including the functioning of container terminals) without the application of innovative, ecological solutions that are indispensable to develop efficient and safe green supply chains.[1] The protection of the natural environment and sustainable development come as the immanent, multi-aspect results of such solutions. This occurs in a simultaneous and independent way, notwithstanding the implementation of other green technologies specifically dedicated to maritime transport (including containerization processes) and related to the implementation of zero-emission (electrical drive) and autonomic (unmanned) container ships (the *Yara Birkeland* feeder project) or container terminals that are neutral to the climate (CTA Hamburg) and come as the prime examples of solutions that have been already implemented in the discussed field.

The previously mentioned argumentation clearly indicates the vast significance of tasks that are performed by maritime container transport for the benefit of the global economy. Hence, the analysis of problems pertaining to the comprehensive identification of factors determining the potential of developing maritime transport, including the broadest spectrum of elements that shape its current and future capabilities, is undoubtedly well-grounded. Understood in such a way, the determinants include economic, organizational, technical, operational, managerial, information, IT and ecological (related to the lowering of anthropogenic impact of transport and its decarbonization) factors.

This approach has allowed the authors to define the thematic range of the following monograph, which includes four parts—Maritime Containerized Transport (the Status Quo of Its Structure and Main Processes), Economic Challenges to Maritime Containerized Transport, Managerial and Operational Challenges to Maritime Containerized Transport and Sustainable Development Challenges to Maritime Containerized Transport—containing 12 chapters that present a scope of the identified determinants/challenges in the most systemic and holistic way.

The first chapter introduces the readers to the problems of containerization by presenting the origins of containerization processes in international trade (forming a bridge between the past and the present). This allows the authors to present a chronology indicating the appearance of pre-container cargo vessels in transport and the place and time when containers (as we know them today) appeared. In order to understand the impact of containerization on modern transport, the implications resulting from the common use of containers in global economy are discussed. Subsequently, this becomes a reason for discussing the scope of the impact exerted by accelerators and systemic/technical/economic barriers that determine the development of containerization in the world (with particular consideration of maritime transport). The chapter also presents an attempt at providing their systemization and hierarchization.

The monograph is focused on considerations pertaining to containerization, namely the use of containers in maritime transport processes. Therefore, it is necessary to define a container as a loading unit in shipping processes (this topic dominates the substantive content of the second chapter). The chapter presents the principles of container standardization (type series), construction issues (with the consideration of technical conditions, endurance and safety conditions) and the taxonomical classification of containers with the consideration of their specific use for shipping diversified types of cargo.

The third chapter presents a systemic approach toward containerization, understood as a part of a bigger entirety, namely the containerized transport system (CTS). The operation of the CTS is based on intermodal connections in containerized cargo transport. Hence, it is necessary to present typical modal solutions, complementary concepts (e.g., combined transport) and basic integration assumptions that underlie the implementation of inter- and multimodal solutions. Integration processes could not be developed without any legislative support and successive implementation of (legal, conventional) regulations referring to all the aspects of the functioning of CTS and other fields related to container turnover. Such regulations include, among others, the International Convention for Safe Containers (CSC) and legal regulations on the transport of containerized hazardous cargo (the International Maritime Dangerous Goods Code).

In the processes of maritime transport, containers are usually shipped by specialized container vessels. Hence, the fourth chapter is focused on problems related to the development of global fleets of container ships. It presents the most common types of container vessels (in accordance with their growing size—up to 24 K TEU included). Based on the statistical data, the structure of the global fleet of container vessels is also presented. Considering the stochastic anomaly (the Covid-19 pandemic), the chapter presents an attempt at evaluating the impact of that global phenomenon on the international trade turnover and the prospects for development and use of the container ship fleet in the world. In Chapter 4, the largest container operators in maritime transport in the world are identified, and the characteristics of the selected shipowners are provided. To provide a full exemplification of the position taken by shipowners on the global market of containerized freight, the economic and organizational aspects (assets, capital, revenues and costs) of the selected container shipowner (Hapag-Lloyd) are presented.

Chapter 5 presents the size and geographical structure of international maritime container transport, focusing on such aspects as the statistics pertaining to the global stock of containers applied in maritime transport, which is determined by the suitability of cargo for containerized shipping in international trade, the volume of flows and directions of global maritime container transport or the volume and geographical structure of global container turnover at seaports. Considering this aspect (the flows of containerized traffic), the problem related to the repositioning of empty containers in international maritime trade becomes of high significance, and it comes as the last question discussed in the fifth chapter. It concludes the considerations of Part 1 dedicated to maritime containerized transport (with particular focus on the current status quo of the structure of maritime containerized transport and the identification of the main processes).

The sixth chapter presents the economic aspects referring to transport and container handling costs in international trade. The chapter also provides an insight into the problems pertaining to the allocation of transport and container handling costs to exporters and importers (Incoterms, Combiterms and RAFTD). It also presents the range of port fees that are charged on maritime container operators for the use of port facilities and container terminals. There are also customs brokerage service fees, freight rates in maritime transport of containers and fees for services provided at container terminals that are charged on exporters and importers of containerized cargo. The considerations presented in the sixth chapter are concluded with the characteristics of the role and significance of indices in maritime container turnover (*World Container Index*—WCI, *Freightos Baltic Index*—FBX, *Shanghai Containerized Freight Index*—SCFI, New ConTex and others).

Chapter 7 presents aspects related to the documentation of processes, providing an insight into documents applied in international maritime transport of containerized cargo. In the chapter, the significance of documents—such as a booking contract, a booking list, a bill of lading, basic types of bills of lading (including liner, direct, multimodal and FIATA bills of lading), a slot-hire agreement and a cargo manifest—is discussed.

Resulting from the real conditions of pragmatics pertaining to shipping processes, the insurance of cargo, containers and vessels in maritime transport is discussed in the eighth chapter. The chapter provides identification of the current circumstances for insurance, focusing on vessel insurance, civil liability insurance pertaining to vessel exploitation and cargo insurance. It concludes the considerations of Part 2 dedicated to economic challenges faced by maritime containerized transport, focused on a holistic approach toward maritime containerized transport costs.

Maritime transport of containerized cargo involves not only maritime shipping but also port operations. Hence, Chapter 9 presents scientific considerations on determining the role of port container terminals in maritime container turnover. It is also necessary to define the notion of a container terminal and its complementary entities (dry ports and container depots). As a result, the chapter presents the functional and organizational characteristics of a maritime container terminal and a definition of the main stakeholders of that system: global operators of maritime container terminals.

In the tenth chapter, the organizational and technical conditions underlying the functioning of port container terminals are presented, starting with the identification of model solutions applied in the field of operating port container terminals (horizontal, vertical, neutral to the climate and multimodal structures—maritime and inland waterway transport, a concept of an IPSI [Improved Port/Ship Interface] terminal). Considering dynamic changes observed in shipping processes, such as the container vessel size maximization, it is important to indicate the influence of these changes on operations performed at terminals, which involve container handling (new concepts of handling mega-container ships at terminals). The optimization potential in the operation of maritime container terminals is characterized by the capabilities of terminal infrastructure and suprastructure. Hence, these problems come as another aspect discussed in the seventh chapter, all the more so that they are directly translated into the quality and range of cargo handling services offered by port container terminals to cargo consigners and consignees (container stuffing and stripping, stowaging containers on board and providing access to the operational system of a terminal for cargo consigners and consignees).

In the context of the optimization processes that have been taking place in maritime transport of containerized cargo, the managerial aspect is also very significant, and it is discussed in Chapter 11. The chapter provides identification of implemented information and IT (telematics) solutions in the containerization processes in maritime transport. Furthermore, the essence of tasks performed by telematics systems in the management of containerization processes is indicated, and monitoring systems of container loading units in maritime transport (of the ECTS [Electronic Container Tracking Service] class) are discussed along with the loading systems applied in maritime transport of containerized cargo (of the vessel planning class). The chapter also presents terminal systems dedicated to maritime transport of containerized cargo (of the TOS [Terminal Operation System] class) and access systems dedicated to cargo consigners and consignees that allow them to access terminal operation systems and that actually reach the key point of process pragmatics (an access system for forwarders, a notification system, OCR). It concludes the considerations of Part 3, dedicated to the managerial and operational challenges to maritime containerized transport, with emphasis on the entire *spectrum* of aspects related to optimization through the implementation of best managerial processes and IT (telematics) solutions/tools.

Chapter 12 is dedicated entirely to the development of sustainable maritime containerized transport. The discussion is started with the identification of ecologistics and requirements of sustainable development in the pragmatics of containerization processes in maritime transport. Furthermore, more aspects such as the origin and formal and legal aspects of emission restrictions in maritime transport, maritime transport as a GHG emitter—an imperative for the implementation of a methodology for counting externalities—and external costs in maritime transport (including containerized maritime transport) as well as legal acts and legislation (GHG protocol, Sustainable Development Goals SDG-17) are mentioned. As a methodological consequence, components of GHG/CO_2 emission and CO_2 equivalent in maritime transport in line with standards, control and verification of emissions generated by maritime transport (MRV CO_2/CO_2e) are considered. This part of Chapter 12 is concluded with the MRV CO_2 identification of the procedures and information about standards, control and verification of emissions generated by maritime container terminals. Finally, organizational, technical and operational factors determining lower emissions generated in the operational field of a maritime container terminal (scope 1 and 2—scope 3 ultimately) as well as a discussion on a maritime container terminal as an element of a low-/zero-emission sea-land logistic chain is presented. It concludes the considerations of Part 4 of the presented monograph, dealing with sustainable development challenges to maritime containerized transport (leading to the conclusions on preparations of maritime containerized transport for further zero-emission requirements).

As presented previously, the organization of the contents clearly indicates that the monograph contributes to the full identification and systemization (hierarchization) of the determinants in the development of maritime transport of containerized cargo. The authors believe that the considerations presented in the monograph cover all the key problems and challenges pertaining to maritime transport and containerization, starting with economic factors, organizational, technical, operational, managerial, information and IT challenges and ending with ecological challenges (in the light of a postulate to lower the anthropogenic impact of maritime transport and its decarbonization). Hence, the purpose of this book is an attempt at providing the broadest (systemic, holistic) approach toward the identification of contemporary challenges and the pragmatics of the functioning of maritime container transport in the reality where dynamic market changes have been caused by the global stochastic phenomena (the impact of the pandemic).

Approached in such a way, the problems discussed in the monograph clearly indicate that it is addressed to a wide range of readers who are interested in theory and practice of containerization and, most of all, to stakeholders of the broadly understood world of maritime economy and maritime transport, also to employees of ports, shipowners of all levels and decision-makers who decide about the structure of the container transport system and international standards implemented in maritime transport. It is also addressed to employees of maritime administration and to students whose fields of study pertain to economics and maritime transport administration.

NOTE

1. Activities in the climate field are the central element of the new EU Green Deal, and the problem of emission control is based on the resolutions of the EU system of emission allowance trading (EU ETS).

Part 1

Maritime Containerized Transport (the Status Quo of Its Structure and Main Processes)

1 The Origins of Containerization Processes in International Trade

1.1 PRE-CONTAINER LOADING UNITS IN TRANSPORT AND THE ARRIVAL OF A CONTAINER

Before the invention of containers, goods used to be loaded onto vessels or other means of transport by hand. This was arduous physical work, often performed by dock workers in haste, when no special attention was paid to safety of goods and cargo. Hence, the goods were often subject to various types of damage. Only some solid packaging could prevent damages to cargo. However, a lot of time had passed before the problem was solved. A container came as a good solution. Containers (as we know them today) were preceded by amphorae, baskets, leather sacks or sacks produced from other materials, barrels and wooden chests (cases) applied in horse and water transport and then in rail transport. Starting in the 19th century, all these receptacles were more and more often replaced by metal containers dedicated for shipping various products. Those containers came as a kind of reusable packaging, which not only protected goods against damage but also facilitated their handling from a horse carriage to a barge or a vessel.

The word *container* comes from *contire*, a Latin word that means *to keep something together*. The English *container* also refers to a receptacle, a canister or a bin. Considering containers, an Anglo-Saxon measurement has been accepted—a foot— instead of a decimal measurement that is more convenient for inhabitants of continental Europe. In the past, particular containers were dedicated to particular groups of goods. For example, amphorae were used for keeping and transporting wine or olive oil; baskets were used for keeping eggs, vegetables and fruit; and chests for keeping various utensils. Today, containers are used for transporting almost all types of cargo (Klose 2012).

The exact time when a container, as such, appeared for the first time is a highly controversial question. However, it is generally assumed that containerization started at the end of the 18th century. The first prototypes of modern containers were introduced to the common use at the turn of the 19th and 20th centuries in Russia, England and France. They were referred to as *lift-vans*, which meant *chests dedicated to be handled by a crane*. As one of the reports provided by the United Nations Conference on Trade and Development (UNCTAD) suggests, in 1906, an American firm, Bowling Green Lift-Van Company, used metal

DOI: 10.1201/9781003330127-3

chests of the size of 18 × 8 × 8ft for transporting general cargo. Those first containers were not adjusted to stacking. In 1929, the Golden Arrow train, which connected London and Paris, started transporting passengers' baggage in containers. In 1933, the International Chamber of Commerce in Paris established the International Container Office in Paris. The institution carried out the research work on the implementation of containerization in cargo shipping, and it promoted containers as convenient transporting units in international trade. In 1934, the International Union of Railways in Paris (*Union Internationale des Chemins de fer*—UIC) developed the first standard for containers in road and rail transport (Sagarra et al. 2009, 15–16).

An official definition of a container was formulated in 1968 by the Technical Committee of the International Organization for Standardization. According to this definition (Neider and Marciniak-Neider 1995, 33):

- a container is a durable unit used for shipping characterized by proper endurance that guarantees its reusability;
- its construction makes it possible for a container to be transported by one or several means of transport, without the necessity to handle the cargo it carries.
- A container is equipped with elements for fastening, maneuvering and handling from one means of transport to another;
- its construction allows users to load and unload the cargo, which is carried inside a container, easily.

A lot of credit for the development of containerization must be given to an American entrepreneur and a visionary Malcolm McLean (1913–2001), who is often referred to as the father of containerization. In 1934, he bought a secondhand truck, and he started a shipping and forwarding company that later on became one of the largest American companies in the sector of road transport, having almost 1,800 vehicles. As he had observed a number of limitations and technical problems related with handling cotton from trucks onto vessels,[1] McLean and K. Tantlinger, an engineer, constructed an intermodal steel container of the size of 35 × 8 × 8ft. It was equipped with special fastenings installed at each corner, and in this way, it could be easily handled by a gantry crane from a port wharf onto a vessel or from a vessel to a wharf (in order to facilitate loading and unloading processes). The invention became a milestone in the development of containerization in maritime transport. The regular shape of containers made it possible to stack them and to use their capacity and loading capacity of vessels in the optimal way. In order to transport containerized cargo by sea, McLean established his own shipping company. He bought the Pan-Atlantic Steamship Company, with 7 C-2 vessels that had been constructed during the Second World War, and then he took over the Waterman Steamship company, making himself an owner of 30 other vessels of C-2 type. At first, his shipping company operated as Pan-Atlantic, and later on as Sea-Land. One of the vessels was converted into a container vessel, which was named the *Ideal X*. She It was equipped with her own onboard gantry crane. On 23rd April 1956, she sailed with 58 aluminum containers from Newark, New Jersey, to Houston, Texas, where 58 trucks with semitrailers were waiting for her to transport the containers further on. McLean calculated that transporting the containerized cargo on that route cost him only USD 0.16 per tonne, whereas transporting it in a traditional way would have cost USD 5.83 per tonne (Valdes 2007).

In 1966, the transoceanic transport of containers between the USA and Europe was launched. The first lot of 226 containers of the capacity of 35 TEU[2] came from New York to Rotterdam and then to Bremen on board and in the holds of the S/S *Fairland*, which belonged to the Sea-Land shipping company controlled by M. McLean. The economic boom of the 1960s in the USA fostered the development of trade with Europe and the launching of regular transatlantic container lines. The use of containers was also forced by more and more frequent port congestions. At the same time, ports on both sides of the Atlantic Ocean took those factors as the reason for adjusting their cargo-handling capabilities and storage capacity to provide services to bigger and bigger container vessels. The Port of Rotterdam was the forerunner in that sector. In the middle of 1967, the first European container terminal was opened there. In the subsequent years, Rotterdam became one of the biggest containers ports in the world. The first ports that received American container vessels were also Bremen and Grangemouth (Scotland) (Szyszko 2021).

It should be emphasized that, at first, transporting containers by sea took place exclusively with the use of vessels that had been adapted for that purpose. However, in the mid-1960s in Australia, the first specialized container vessel was launched. In the course of time, container vessels almost entirely ousted classic general cargo vessels from sea routes. Simultaneously, port-to-port container shipping was largely transformed into door-to-door shipping.

The Vietnam War significantly affected the acceleration of containerization in maritime transport. When, in the mid-1960s, the USA was militarily engaged in Vietnam on a large scale, which required massive and systematic supplies of military equipment, food and other necessities delivered by sea to American soldiers, the Pentagon struggled with some huge logistics problems that it could not solve solely by itself. McLean efficiently engaged himself into solving those problems. He decided to rely on containers and vessels for shipping them to provide supplies to the American troops in Vietnam. When a problem of empty journeys of container vessels on the way back from Vietnam to the USA appeared, McLean suggested that the vessels should visit Japan and then other ports of Southeast Asia and take cargo that was going to be exported to the USA (Valdes 2007). This provided some new opportunities in the field related to the development of international trade between Asian countries with large resources of cheap labor and rich countries of the West where labor costs were high. Owing to the fact that the introduction of containers resulted in the lowering of transport costs,[3] it became economically justified to move production of labor-consuming products to China, India and other Asian countries offering low labor costs. This fact became a strong stimulator to global trade and a catalyzer for containerization processes (especially in maritime transport).

1.2   IMPLICATIONS CAUSED BY THE IMPLEMENTATION OF CONTAINERS TO THE WORLD'S ECONOMY

Containerization is a process of using containers that meet international standards in all modes of transport and adjusting them to the entire transport system. Thanks to such advantages as simple and relatively cheap construction, a large capacity, the possibility to be handled quickly from one means of transport to another with the use of mechanical (and largely automatic) devices and other qualities, containers have become a strong drive to the dynamic development of containerization in the world.

Containerization has radically transformed cargo shipping in all the modes of transport, but it can be observed—perhaps in the most conspicuous way—in maritime transport. Considering the intensity of changes in transport, the arrival of containers is nowadays referred to as its third revolution. The significance of that event is therefore comparable to the invention of the wheel (Mesopotamia, 3500 BC), which was the beginning of the first revolution in transport. It can be also compared to the use of a steam engine by James Watt in 1763 which was applied as a drive for various means of transport and which came as its second revolution (Neider and Marciniak-Neider 1997, 24). Containerization has become an important factor in the development of global trade and globalization processes. The development of commodity trade in the world generates additional demand for transporting containers by sea, and it encourages shipowners to enlarge their fleets of container vessels.

Compared to shipping goods by classic general cargo vessels, transporting cargo in containers has a lot of advantages, both for shippers and for carriers of cargo. The main advantages for cargo shippers involve reduced transport costs. It results from the shortening of the time of the entire shipping cycle, starting from the moment of loading goods on the means of transport and ending with unloading goods at their destination. The advantages also include savings related to packaging and lower risk of cargo damage or loss. These factors are in turn translated into lower insurance rates. Furthermore, containerization allows shippers to improve the timeliness of deliveries and to shorten the freezing time of the capital that is tied up in transported goods.

Advantages for cargo carriers refer to the fact that heterogeneous cargo loaded into a container becomes homogeneous cargo, and therefore, it can be loaded onto a vessel more easily and quickly; it also allows carriers to use the loading capacity of vessels in a better way. Containerization allows carriers to shorten the time when various means of transport stay at ports or any other transport hubs and—in this way—to reduce costs of handling those means of transport. The relation of the time when traditional general cargo vessels stay at ports to the time they spend in their sea journeys is 50:50 on average; for container vessels, this relation is 20:80. This fact has improved the efficiency of vessel operation because vessels earn money when they are at sea. Handling classic general cargo vessels at ports in times when a lot of operations were performed manually could take weeks—now loading and unloading a container vessel takes one or two days.[4] Moreover, improving the safety of goods by transporting them in containers allows carriers to reduce costs related to the shipowners' liability for damaged goods. In considering a military aspect, the advantage of using containers includes protection provided to the transported military equipment and the possibility to keep that equipment transported in secret.

The use of containers in maritime transport has significantly changed the structure of the global maritime fleet, the character of seaports and people who work there. The importance of traditional general cargo vessels has been decreased in favor of container vessels, which now transport over 90% of all general cargo in international maritime trade. Loading and unloading container vessels become more and more automated, and this fact entails a decrease in employment at ports. In loading and unloading a container vessel, some specialized equipment is needed— reach stackers, various types of cranes and forklift trucks, which are all operated by qualified staff. Specialists are also needed for stowing containers on board. These employees have largely replaced traditional dock workers who used to perform heavy physical work related to the turnover of general cargo at ports. In relation to the ongoing digitalization of the whole port economy, container terminals need more and more computer programmers and IT specialists. Considering the fact that container vessels need very short time to stay at ports, fewer kilometers of port wharves are required, and more storage yards for containers are needed instead. At the same time, however, the use of containers in international trade results in a threat that they can be used to hide drugs or to smuggle other products that are forbidden in international trade. The fight against this type of crime is very difficult because of huge numbers of containers that are transported from abroad and because there is no possibility to check every single container. The problem in container turnover is also to find cargo that can be loaded into containers that are backhauled to their base depots to eliminate empty runs.

Containerization has also caused significant changes to railway and road transport: trucks and railway flatcars adjusted to transport containers have appeared. It has also been necessary to construct roads meeting more stringent technical specifications. New freight hubs have been constructed, taking the form of container terminals, and places of customs clearance have been shifted from the state boarders to premises that belong to consigners of goods, containers have been introduced to the places where goods are manufactured, etc.

To sum up, the process of using intermodal containers has radically changed the operation of shipping general cargo by sea, and it has also significantly affected containerization in inland waterway transport. The process of containerization has resulted in the shortening of time when vessels (seagoing ships and barges) stay at ports and also in a more efficient use of loading space on ships and port storage yards by multilayered stacking of containers. The innovations that have been implemented in shipping have already minimized the risk of cargo damage during shipping and cargo-handling processes. Furthermore, they have also reduced cargo-handling costs (Ficoń 2013, 321).

However, the dynamic market environment, ongoing internationalization of trade relations, globalization of international trade and the specific role of maritime transport in international trade come as the key factors (determinants) in the further development of containerization.

1.3   DETERMINANTS OF THE DEVELOPMENT OF CONTAINERIZATION IN THE WORLD WITH THE PARTICULAR CONSIDERATION OF MARITIME TRANSPORT

The standardization of the sizes of containers and of their maximal gross weight has a fundamental significance to the expansion of containers in all the modes of transport. The first series of container standardization regulations was provided by the International Organization for Standardization (ISO) in 1968. It covered the most popular, universal 20ft containers of the capacity equal to 1 TEU (Twenty Equivalent Unit—20 × 8 × 8ft) and 40ft containers of the capacity equal to 2 TEU, which are also referred to as FEU (Forty Equivalent Unit—40 × 8 × 8ft). After that, the subsequent series of standardization regulations took place in the field of the container use. In the 21st century, 45ft containers have been introduced into use at a larger scale. When fully loaded, 45ft containers allow users to achieve higher profitability in transport than 20ft and 40ft containers (Bartosiewicz 2013, 132). The second important step in the development of containerization was the establishment of Intercontainer, an international partnership. It was established in 1967 by the railway management boards of 11 European countries to organize container rail transport (Krasucki and Neider 1986, 24).

The earliest container shipping by sea at the largest scale was launched in the North Atlantic, between ports in the USA and ports in the developed European countries. Then it spread to the Pacific Ocean, connecting North America, Australia and Southeast Asia. In the 1970s, containers were applied to transport cargo in trade between old European metropolitan areas and their former colonies. Subsequently, a dynamic development of cargo shipping by rail and by road between seaports and their economic hinterlands was observed. In 1973, large container vessels appeared for the first time along oceanic routes connecting ports in the Far East with ports in Europe. In 1984, the first global container line was launched. In the course of time, containerized cargo transport has dominated the entire shipping of general cargo by sea. Today, their share in global shipping of general cargo by sea exceeds the level of 90%.

A very fast increase in container transport was observed in the 1970s. It was then when the first container vessels were constructed as ships dedicated exclusively for shipping containerized cargo. In 1970, containers of the total loading capacity of 3.2 million TEU were transported by sea, but in 1980, that number was increased up to 18.7 million TEU. The development of containerization in maritime transport in the years 1980–2019 is presented in Table 1.1.

Considering the years 2000–2019 only, it is possible to observe that the volume of containerized cargo shipping in tonnes was increased more than three times. It was growing much faster than global transport of cargo by sea, and as a result, its market share was increased from 10% in 2000 to 17.7% in 2019. During the discussed period of time, the assortment of goods transported in containers was extended, and next to fruit and vegetables, clothes and packaged household items, it also included sacked goods, machinery, equipment and steel goods. It was possible because of special-purpose containers that were introduced to transport: reefer containers, insulated containers, tank containers, dry cargo containers, etc.

Containerization was the most successful process in developed countries, where labor costs were high. At first, the excess of cheap labor in developing countries and high

TABLE 1.1

Containerized Cargo Transport in Comparison to Global Shipping in Maritime Transport in the Years 1980–2019

Years	Cargo shipping by sea (given in millions of tonnes)	Containerized cargo shipping (given in millions of tonnes)	The share of container shipping (given in %)
1980	3,704	102	2.8
1985	3,330	152	4.6
1990	4,008	234	5.8
1995	4,651	371	8.0
2000	5,984	598	10.0
2005	7,109	1,001	14.1
2006	7,700	1,092	14.2
2007	8,034	1,215	15.1
2008	8,229	1,272	15.5
2009	7,859	1,134	14.4
2010	8,409	1,291	15.4
2011	8,785	1,411	16.1
2012	9,197	1,458	15.9
2013	9,514	1,532	16.1
2014	9,843	1,622	16.5
2015	10,024	1,660	16.6
2016	10,289	1,734	16.9
2017	10,702	1,834	17.1
2018	11,029	1,927	17.5
2019	11,078	1,966	17.7

Source: Data from Statista (2021g, 2021f), UNCTAD Stat (2020c).

investment costs at container terminals did not encourage development of containerization. Also, there were no conditions for the development of containerization in the countries that exported mainly raw materials. They imported machinery and consumer goods suitable for container transport, but the strong imbalance in container turnover in both directions would have negatively affected the efficiency of container shipping. In the course of time, the situation was improved because—as a result of dividing the value chain—developing countries became important producers of labor-consuming goods that were going to be exported to developed countries, and the foreign capital started to show more interest in investment into container terminals (Grzybowski et al. 1997, 181–182).

The dynamic development of containerization would not have been possible but for implementation of big container vessels to maritime transport and an increase in their number. The number of container vessels (of the capacity of over 300 GT)[5] was increased from 750 ships in 1980 to 4,677 in 2010 and 5,301 in 2019. The capacity of the global fleet of container vessels was increased from 11 million DWT in 1980 to 64 million DWT in 2000 and 266 million DWT in 2019 (see Table 1.2).

TABLE 1.2

The Capacity of the Global Fleet of Container Vessels in the Years 1980–2019 (Given in Millions of DWT)

1980	1985	1990	1995	2000	2005	2010	2015	2016	2017	2018	2019
11	20	26	44	64	98	169	228	244	246	253	266

Source: Data from Statista (2021a).

The size of container vessels was also systematically enlarged, which allowed interested parties to take advantage of the economies of scale. However, for many years, the dimensions of the Panama Canal had been a serious limitation, until 2015, when it was eventually widened and made accessible for vessels with a loading capacity of 12,000 TEU. At the end of 2020, the largest container vessel was the HMM *Algeciras* at 400m in length, 61m in beam and the capacity of 24,000 TEU. She was built at the Daewoo and Samsung shipyard in 2020 for the Korean HMM shipowner. Now she is used for transporting containers between Southeast Asia and Northern Europe (Wikipedia.org 2019). When calculated per one fully loaded container, transport costs by mega-container vessels are lower in comparison to smaller vessels; however, the weak point of these mega-ships is that they can call only at a limited number of ports that meet technical parameters required to handle such huge vessels. This fact can become one of the major barriers to the further enlarging of the size of container vessels in the world.

Simultaneously with development of container transport, the demand for containers was growing, and as a result, an increase in the global stock of containers was observed. In 1965, it was 80,000 TEU, but in 1996, it was over 10 million TEU. The container stock was dominated by 20ft general-purpose containers, but the number of special-purpose containers was also systematically increased, mainly reefer containers and open top containers. In the mid-1990s, about 50% of all containers were owned by leasing companies (Pawlik 1999, 38). In 2010, the total number of containers counted in TEU was increased up to 28 million. In 2008, the share of 20ft containers was decreased to 30% in favor of 40ft containers. In the years 1980–2010, the share of sea carriers in the ownership of the global container stock increased from 53% to 55%, at the expense of leasing companies. The share of other container operators was small, and it equaled only 1.3%. Those operators were mainly railway carriers (Marciniak-Neider and Neider, eds. 2014, 309).

In order to provide a synthetic approach toward all the potential determinants of the further development of containerization processes in maritime transport, the following key factors (indicating the growth limits) should be mentioned:

- capabilities for further growth of international trade volume (in the context of economic and geopolitical relations, implemented terms of trade and disturbances caused by the pandemic)—limits to international trade

- capabilities for further development of container vessels (in relation to their size given in TEU and the limits to the economics of their use)—limits to container ship capacity[6]
- capabilities for the supply of new containers (in relation to the capabilities of the industry responsible for container construction)—limits to the container volume
- capabilities for further development of terminal infrastructure (in relation to the size, expansion, levels of competitiveness and operational efficiency of container terminals)—limits to container terminal capacity[7]
- capabilities for further development and capacity of maritime transport infrastructure (in relation to navigation sea routes and canals, especially the Suez Canal and the Panama Canal)—limits to the capacity of sea communication routes[8]
- the level of safety and security of maritime and terminal/port operations (including both these aspects: safety and security and also safe container turnover)—limits to maritime (container) safety and security[9]
- capabilities for further lowering of maritime transport anthropopressure (in relation to a decrease in emission of greenhouse gases—GHG, CO_2 and the general lowering of external effects)—limits to the internal costs of container transport
- capabilities for further digitalization and computerization of maritime and container operations (in relation to automation, robotization, autonomation, implementation of IT systems supporting management, the use of database systems, artificial, virtual and augmented reality)—limits to information management

The division of global relations and economic exchange into so called "triad regions"—namely political and economic conditions for trade between North America, Asia and Europe—may come as an example of the impact exerted by international relationships and geopolitical conditions on the shape of maritime container connections. The current tensions observed in the trade between the USA and China perfectly illustrate the fragile character of these relations and their vulnerability to destabilization. The scope of international trade between global stakeholders is presented in Figure 1.1.

A good example illustrating infrastructural limitations to the development of containerization in maritime (including container) transport is a situation that took place in the Suez Canal on 23rd March 2021.[10] The canal was blocked by a mega-container vessel, the *Ever Given* (Evergreen shipping company). The vessel, in her 399.994m total length, 58.80m moulded beam and 16.00m draught, ran aground across the southern single-lane section of the canal after she had lost her maneuverability in strong wind and a dust storm.

It took six days to free the ship. In the meantime, on both ends of the canal (Port Said and Suez), several hundred vessels, including large container ships, tankers with oil and gas and bulk carriers with grain, formed one of the largest navigation congestions ever.[11] Such a gigantic congestion had never happened before in that area. The scale of the blockage is presented in Figure 1.2.

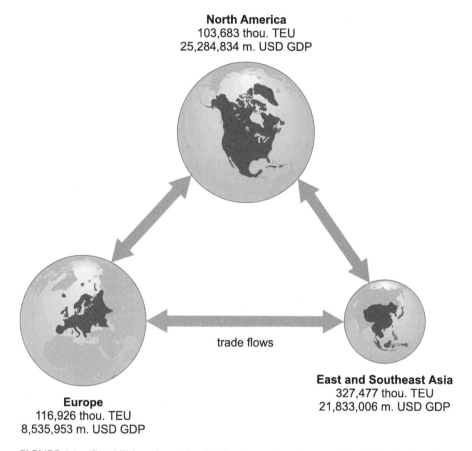

North America
103,683 thou. TEU
25,284,834 m. USD GDP

trade flows

East and Southeast Asia
327,477 thou. TEU
21,833,006 m. USD GDP

Europe
116,926 thou. TEU
8,535,953 m. USD GDP

FIGURE 1.1   Capabilities of containerization in maritime transport in relation to the structure of international trade in the "triad" approach (the status as in 2020).

Source: Adapted from Pearson Education Inc. www.pearson.com/us/ (accessed: 10th July 2018).

Considering the general lines of changes taking place in transport and presenting their further synthesis, it should be assumed that in the subsequent years, the competitiveness of maritime transport of containerized cargo and capabilities of its further development will be decided only by two groups of factors (Miler 2016b, 148–149):

- digitalization, automation and autonomation
- protection of natural environment[12]

Nowadays, in the global world, it is impossible to maintain competitiveness and capabilities of development without innovative digital tools that are indispensable to develop an efficient and safe supply chain. The protection of the natural environment and sustainable development are inseparable, multi-aspect results of their use, regardless of green technologies that are specially dedicated to the sector of maritime transport (including containerization processes).

FIGURE 1.2   The congestion after the blockage of the Suez Canal (status as for 25th March 2021).

Source: Reprinted with permission of Fotini Tseroni, Content Writer, Marine Traffic/marinetraffic.com (accessed: 25th March 2021).

One of the systems in the digital world of logistics and maritime (container) transport is the Port Community System (PCS).[13] This is a modern, neutral and safe platform that has not been dominated by other systems, and at the same time, it is compatible with those systems, providing a proper, common and integrated format of information that flows every day to the selected users (under logistics processes, the EDI standard is applied—Electronic Data Interchange). Seaports also use automation processes more and more extensively. The Altenwerder, an automated container terminal (CTA), has been operated in Hamburg for many years. It handles cargo without any direct participation of workers and with the use of autonomic (driverless) vehicles (AGV—*Automated Guided Vehicle* and *Double Rail Mounted Gantry*—DRMG). This is the first sea container terminal that is neutral in terms of its impact on the climate. The International Container Terminal Services (ICTSI), a Philippine port group, opened an autonomous container terminal, Victoria, in Melbourne, Australia. The biggest port in the world, Shanghai, which in 2017 handled over 40 million TEU, launched a fully autonomous terminal, Yangshan Deep Water Port (YDWP), of the capacity of 6.3 million TEU per year.[14] More and more intensely, digitalization is extended onto the operational aspects of the functioning of maritime transport, especially on the aspect of mutual interactions between maritime transport operators and their contracting parties (more broadly, stakeholders of the logistic supply chain)—for example, in the form of blockchain technology.[15]

Regardless of the mentioned processes, some other innovative projects have been implemented in the field of the protection of the natural environment and zero-emission autonomous transport (autonomous seagoing vessels with electric drives). Maritime transport is changing under the influence of IT/ICT (Information Technology/Information and Communication Technology). It is more and more often possible to observe that the value (as well as the failure) of the human factor has been decreased through the use of augmented reality (AR), virtual reality (VR), artificial intelligence (AI) (CargoX 2017) and 5G networks, which enable common use of smart cameras and computer vision systems for full automation or digital twin technologies that are used for the visualization of 3D processes in real time (Tideworks 2021a).

The implementation of autonomous vessels is a solution that comes as a step into the future for the maritime sector.[16] Yara, a Norwegian company that produces artificial fertilizers, and Kongsberg, a company that develops solutions in the field of maritime technology and telematics, have combined their efforts to implement an exceptional project. A feeder line from the factory to the local port has been handled in a fully autonomous way since the beginning of 2021. The first autonomous vessel, the *Yara Birkeland*, a small container vessel with a capacity equaling 120 TEU, propelled by an electric drive, was launched at the end of 2020.

The concept of an autonomous vessel is based on the assumption that advanced telematics devices can replace work performed by watchkeeping officers, and additionally, they can be improved by the operation of infrared systems in order to obtain better performance. With the support of the independent interpretation of data obtained from devices such as sonars, radars and LIDARs,[17] which have so far been operated by people, an autonomous vessel will require only some occasional verification from a remote operation center (an operation/management center).

The lack of a crew will contribute to the reduction of costs related to the elimination of infrastructure and cabins indispensable for people working on a ship. It will also provide more loading space, which will translate into higher profits from the operation of a vessel. Another advantage is the optimal adjustment of the vessel construction to loading/unloading procedures. As estimated for the *Yara Birkeland* project, thanks to the introduction of that new vessel into service, 40,000 fewer vehicles (tractor units with container semitrailers) will travel along the route between Porsgrunn (Heraya) and Larvik (Norway) per year (Chip.pl 2017).

The essence of containerization processes understood in such a way (and also one of the key determinants of their further development) is a loading unit, namely a container.

NOTES

1. At that time, loading a whole vehicle with its cargo on board was impossible because it would have taken too much space. However, in a conceptual aspect, it was the beginning of the development of containerization.
2. *Twenty Equivalent Unit* is the equivalent of a 20ft container.
3. For example, considering the import of shirts from China to Europe, the share of the transport costs in the selling price does not exceed 1%, and the transport cost of a pair of jeans on that route is approximately (USD) 30 cents.
4. See more in Brzozowski (2012).
5. The gross tonnage (GT, G.T. or gt.) is the measurement of gross capacity (tonnage) of commercial vessels. Despite its name, gross tonnage does not indicate mass—it indicates the capacity (defined with the use of special equations and measurements) of all the spaces inside the hull and the superstructure of a ship, excluding ballast tanks. Until 1994, gross tonnage used to be given in gross registered tonnes. A registered tonne was equal to the capacity of 100 ft^3, which is 2.83 m^3. Based on the Convention of 1969, in 1982, the way of measuring capacity was changed, and the capacity of vessels was given in bare (dimensionless) units as gross tonnage (GT) and net tonnage (NT); in 1994, the principle of using registered tonnes was cancelled. Vessels whose gross tonnage and net tonnage have been determined in accordance with the convention are given International Tonnage Certificates (since 1969).
6. Technical literature and experts' opinions define the limit at the level of 50 thousand TEU; see more at www.porttechnology.org/news/mckinsey_report_vessels_to_reach_50000_teu_by_2067/.
7. Technical literature and experts' opinions define the limit at the level of 80 million TEU at one terminal; see more at www.porttechnology.org/news/mckinsey_report_vessels_to_reach_50000_teu_by_2067/.
8. The critical role of sea canals comes as a determinant of the optimization of navigation routes; despite the enlargement and modernization of both key canals (the Suez Canal and the Panama Canal), these narrow passages are the main factors limiting the number, size and type of ships that can pass through the canals safely. In this way, they also determine the capabilities of maritime transport infrastructure.
9. The level of safety refers to factors that are technical and exploitative in nature (technical status, navigation aids, nautical safety of sailing); security is determined by factors related to anthropopressure—piracy, terrorism, sabotage, etc. See more in Miler (2016a).
10. Every day, 50 vessels go through the Suez Canal, which is 193 km long. They transport approximately 25 million tonnes of goods, which account for about 30% of the global

container stock and about 12% of the global trade involving all types goods, including oil and oil derivatives. If passing through the Suez Canal is impossible, it is necessary to circumnavigate Africa and prolonging the journey to Europe by seven days on average.

11. At its peak, the congestion was formed by 453 vessels; the loss for global economy and the Egyptian economy reached the level of hundreds of USD million per day.

12. Generally, in maritime (containerized) transport, all interested parties look for solutions that foster the decarbonization of navigation, the development of innovative technologies and favorable solutions in the construction of low- and zero-emission vessels, the ecological reconstruction of vessels, the use of alternative and renewable types of fuel and respective conditions referring to changes in norms, legal regulations and good practice in maritime transport. These solutions must be compliant with the expectations of the Green Deal.

13. It is operated in 33 ports in the world (status as of the end of 2017). In Europe, it is operated in Rotterdam as Portbase, in Hamburg as Dakosy, in Antwerp as the system of the Port of Antwerp and as a *dbh* system in Bremen and Bremerhaven. The authorities of all big seaports have been considering the idea of implementing the PCS.

14. It allows operators to save a lot on human labor (and to lower the risk of errors in the man-object (technology) -environment (MTE) relation), to significantly reduce costs and the negative influence of cargo handling operations on natural environment, and in the particular case of YDWP, to lower CO_2 emission by 10%.

15. An innovative platform, CargoX Smart Bill of Lading, allows companies to issue and to handle original bills of lading in the Ethereum blockchain network, facilitating the digital transfer of the ownership of bills of lading and related assets from the consigner to the consignee, shipping agent, shipowner or forwarder. Ethereum is a decentralized platform in a cloud that handles "intelligent" contracts on the basis of applications that operate precisely in the way they have been programmed and, importantly, without the interference of any third parties. Bills of lading and related documents are accessible in the cloud of data in their encrypted forms. The combination of those tools has allowed interested parties to develop the first fully digital and safe solution to handle maritime bills of lading (see more at https://log24.pl/news/blockchain-w-transporcie-morskim/).

16. The idea of introducing autonomous vessels into maritime navigation has got enormous potential. The additional reduction of costs related to the maintenance of a vessel would emphasize even more intensely the advantage of such vessels. It would also improve the competitiveness of short-sea transport, which competes with its main rivals: rail and road modes of transport. The project of autonomous vessels perfectly matches the rhetoric about the automation of the supply chain, and it should be developed as a solution of the future. However, it involves as many advantages as disadvantages, and all effort should be made to minimize the hazard. Economic advantages are undeniable. Still, the navigation of autonomous vessels at sea may turn out to be problematic. The COLReg regulations would have to be changed in a fundamental way in order to allow vessel navigating officers and program authors to know how their vessels should respond to each other. The ARPA radar systems are able to suggest navigators how to avoid a collision, thanks to computational algorithms. So far, navigators have been responsible for making the ultimate decisions about the change of the course, considering not only radar observations but also all the external factors affecting the movement of the vessel. Based entirely on a radar, the navigation of autonomous vessels seems to involve quite a big risk. Maritime administration should consider the possibility of such situations and prepare proper regulations of maritime law and regulations of maritime traffic. The introduction of autonomous vessels to the commercial fleet shall come as a huge challenge also to classification

associations, which will have to prepare relevant safety and technological standards. The minimal safety limits will be defined by contemporary vessels. In order to perform active duties, unmanned vessels will have to be as safe as those that navigate the sea at present. The IMO provides highly precise regulations referring to the vessel equipment and reliability. Apart from the aspects of safe navigation and applied technologies, autonomous vessels will have to face the problem of internet piracy. For the constructors of autonomous vessels, the key question will be to secure all the communication access to prevent any attacks against economic and technological security. Apart from technological development, it will be necessary to develop relevant standards and requirements in the field of maritime safety and security management.

17. LIDAR—Light Detection and Ranging—is a device that operates on a principle similar to a radar, but it uses light instead of microwaves. The device is characterized by a high resolution, and it is applied in procedures related to the numeric modelling of a terrain, among others. It provides a numeric, discrete (point) representation of the topographic elevation of the terrain surface with an interpolation algorithm (also in 3D technology).

2 A Container as a Loading Unit

2.1 THE DEFINITION AND PRINCIPLES OF CONTAINER STANDARDIZATION (TYPE SERIES)

A container is an appliance for shipping purposes that is characterized by many specific constructional and functional features. A common definition of a container was introduced by the International Organization for Standardization—ISO) responsible for the implementation of technical standards (Standard 668 of 1968) for container sizes and construction. In accordance with that definition, because of its robust structure, a container should make it possible for cargo to be transported with the use of several means of transport and to be easily handled (Grzybowski et al. 1997, 74; Nierzwicki et al. 1997, 18; Jakowski 2017, 14). An important condition for the implementation of containerization was to define and to adopt some standards related to the standardized dimensions of containers (however, it applied not only to the ISO standards but also to standardization and compatibility of transport units in other branches).

The implementation of a document that regulates standards for the construction and principles for safe container handling by the International Maritime Organization (IMO) has been a significant moment for container transport by seagoing vessels. The document is the *International Convention for Safe Containers 72 (CSC)* of 1972. The convention adopts a definition of a container that is very similar to the previous one suggested by the ISO. In accordance with the convention, a container is a device for shipping that meets the following requirements (Art.II,1.) (*Journal of Laws* 1972, 11):

- is durable and reusable;
- allows operators to transport cargo by several means of transport without the necessity of transshipment handling;
- can be fixed and handled and is adequately equipped with corner castings;[1]
- the surface between four container corner castings should be $14m^2$ or $7m^2$.

The convention refers to containers transported by seagoing ships and by land on container semitrailers and railway cars. Vehicles and cargo packaging are not considered containers. The CSC regulations are not applicable to air freight containers, which are not applied in maritime transport.

In 1997, the International Labour Organization and the United Nations Economic Commission for Europe introduced the *Guidelines for Packing of Cargo Transport Units*—CTUs (IMO/ILO/UN ECE Guidelines 1997, 4). The document includes transport procedures referring to container units, semitrailers, vehicles, railway cars and other non-bulk cargo. An increase in shipping of unitized general cargo (containers, semitrailers and trucks) has resulted in the necessity to update the guidelines referring to proper handling of such cargo when transported.

DOI: 10.1201/9781003330127-4

In 2014, the IMO/ILO/UNECE *Code of Practice for Packing Cargo Transport Units*—commonly referred to as the CTU Code—was published. The code repeats the provisions of the document of 1997, and it provides a definition of a freight container. According to the Code, "a freight container means an article of transport equipment that is of a permanent character and accordingly strong enough to be suitable for repeated use" (IMO/ILO/UNECE Code of Practice 2014, 10). This type of device should be constructed specifically for the implementation of transport processes with the use of various means of transport, without the necessity of stripping and handling the particular loading units. The definition of a container also includes its necessary equipment that allows operators to transport cargo in a safe way and that should be compliant with the CSC Convention of 1972. The CTU Code clearly distinguishes a container from other loading units, however, with the consideration of a possibility of transporting containers with the use of various types of chassis (road and railway transport).

The parameters of intermodal containers undergo normalization under the standardization processes implemented by the ISO. One foot has been adopted as a basic module—namely 304.8 mm. The maximal mass of containers has been limited with the consideration of their carrying capacity and lifting capacity of cargo handling devices (Markusik 2013, 37–41).

To sum up the previously mentioned interpretations, a container is a durable and reusable device for shipping that can be transported by one or several means of transport without the necessity of handling and transshipping the cargo it carries (PN-ISO 830 1999). Moreover, a container should allow operators to handle cargo easily during transshipment operations from one means of transport onto another.

A comparison of various definitions of a container provided in several documents allows the authors to state that one and the same type of device for shipping that meets the required technical standards is discussed here (Table 2.1).

In expert literature, documents and digital sources, it is possible to find various definitions, such as an ISO container, a large-size container, a loading container or a maritime container. In most cases, these terms refer to standardized containers introduced into use by the ISO. It is also possible to find containers that meet other technical standards and that can be used for domestic transport (e.g., in the USA). However, it should be noted that ISO standardized containers are the most common in maritime and road transport in the world. The widespread use of containers in international transport should be credited mainly to the ISO and its consistent activities in the field of implementing technical standards for containers. The standards refer, first of all, to external and internal dimensions of containers (length, width and height), their total permissible weight, location of standardized corner castings, location of forklift pockets for container stackers and gooseneck tunnels for port terminal tractors (Grzybowski et al. 1997, 74).

Considering the previously mentioned definitions and the requirements of this monograph, it is possible to state that a container is an auxiliary device dedicated and properly adjusted for transporting various types of cargo (general cargo, dry and liquid bulk cargo and hazardous cargo) by various means of transport, using various transport branches, without the necessity of unpacking cargo in transhipment.

A basic intermodal (ISO) container applied in shipping and logistics processes (also used as a kind of "a settlement unit") is a hardtop steel container 1C, 20ft, which is a TEU of 20,320 kg of gross weight, 2,300 kg of tare weight, 18,020 kg of loading capacity, 13.5 m^2 of loading surface and a cubic capacity of 30 m^3.

TABLE 2.1

A Comparison of Container Features Listed in Various Definitions

Container features	ISO Standard 668 of 1968	CSC Convention of 1972	Guidelines for CTUs of 1997	Code of practice for CTUs of 2014
Stable and robust structure	Yes	Yes	Yes	Yes
Reusable	Yes	Yes	Yes	Yes
Used in shipping processes	Yes	Yes	Yes	Yes
Can be transported by various means of transport	Yes	Yes	Yes	Yes
Used in maritime transport (a possibility of handling cargo ship to ship, ship to shore)	Yes	Yes	Yes	Yes
Used in shipping processes from a consignor to a consignee			Yes	
During transportation, it is not necessary to handle the cargo inside a container	Yes	Yes	Yes	Yes
Standardized external and internal dimensions	Yes	Yes	Yes	Yes
Equipped with fittings allowing operators to handle a container easily (loading and unloading) onto means of transport—corner castings	Yes	Yes	Yes	Yes
Easy for stuffing and stripping	Yes			Yes
Suitable for transporting receptacles, packaging and unitized cargo			Yes	
Equipped with internal fitting to provide safe cargo shipping				Yes
Internal capacity of 1m³ or higher	Yes		Yes	

Source: Data from MSC Mediterranean Shipping Company (2021, 11), IMO/ILO/UN ECE Guidelines (1997, 4), IMO/ILO/UN ECE Code of Practice (2014, 100, Polish Register of Shipping PRS (2016, 5).

General-purpose containers are equipped with a front door, which is 2,285 mm wide and 2,135 mm high. The open door must be secured, and the rotation of the door leaves must be of at least 260°. Other significant features of a container include its durable and rigid structure as well as its tightness against weather conditions. A container should transfer the strain of dynamic forces that are induced by the acceleration up to 2g in the horizontal plane (Markusik 2013, 67).

To transport light cargo taking up much space and of the height that does not allow operators to pack it into a standardized container, 20/40ft high cube containers are used. Their height is extended to 9ft, 6 in (2,895 mm). As a result, a high cube container offers an additional cubic capacity of 10m³ (for a 40ft container) (SpedCont 2014).

The standardization of containers is a continuous process that takes place with the consideration of the current needs in the field of transport or solutions that have been already applied. The norms referring to the standardization of containers appeared in 1968, and then the first type series including units 5, 6.5, 10, 20, 30, and 40ft of length and 8ft of height was introduced (Figure 2.1).

Since the implementation of the first series of containers, there have been seven subsequent standard versions. The latest version was published in 2020 (Norm no. 668:2020). The particular versions of the standards resulted from the improvement of the container systems that had to meet the increasing demand for shipping services in the field of containerized cargo. Hence, containers that did not meet the expectations were withdrawn, and new types were introduced to fulfill shipping requirements in a better way. In the course of time, the standard implemented by the ISO has been adopted in maritime, land and inland water modes of transport. The functioning container system was initially designed for the purposes of general cargo shipping, but later on, containers were developed to transport almost all groups of cargo. An early type series of containers was based on the multiplication or divisibility of the dimensions of a 20ft container, which became the measure for container loading capacity, and it was referred to as the equivalent of a 20ft container capacity TEU.

Considering the limited capacity of 8ft high containers, 8.6ft and 9.6ft containers have been introduced as they offer more possibilities for strapping the transported

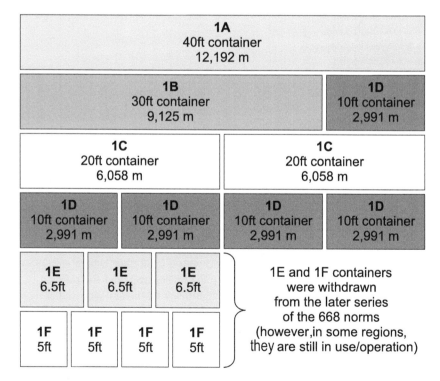

FIGURE 2.1   The type series of ISO containers introduced in 1968.

Source: Data from IMO/ILO/UN ECE Code of Practice (2014, 10), Polish Register of Shipping PRS (2016, 5).

cargo. In 2005, a 45ft container was incorporated into the ISO system. It allows operators to increase loading possibilities up to 32 Euro-pallets, while 40ft containers offer the possibility to transport 25 Euro-pallets (Wiśnicki 2010, 388). The precise container parameters applied in maritime transport are presented in Annex 1. The basic parameters of ISO-type containers are presented in Table 2.2.

TABLE 2.2
The Parameters of ISO-Type Containers

Container type	Linear external dimensions (given in mm)			Linear internal dimensions (given in mm)			Capacity (given in m³)	Gross weight (given in kg)
	Length	Width	Height	Length	Width	Height		
1A (40′)	12,192	2,438 (8′)	2,438 (8′)	11,998	2,330	2,197	61	30,480
1AA (40′)	12,192		2,591 (8′6″)	11,998		2,350	67.2	
1AAA (40′)	12,192		2,896 (9′6″)	11,998		2,655	76.6	
1B (30′)	9,125		2,438 (8′)	8,931		2,197	45.7	
1BB (30′)	9,125		2,591 (8′6″)	8,931		2,350	48.9	
1BBB (30′)	9,125		2,896 (9′6″)	8,931		2,655		
1C (20′)	6,058		2,438 (8′)	5,867		2,197	30	
1CC (20′)	6,058		2,591 (8′6″)	5,867		2,350	32.1–33.2	
1CCC (20′)	6,058		2,896 (9′6″)	5,867		2,655		
1D (10′)	2,991		2,438 (8′)	2,802		2,197	14.3	10,160
1EE (45′)	13,716		2,591 (8′6″)	13,542		2,350	81	30,480
1EEE (45′)	13,716		2,896 (9′6″)	13,542		2,655	86	
1AX (40′)	12,192			11,998			–	–
1BX (30′)	9,125		below 2,438 (up to 8′)	8,931			–	–
1CX (20′)	6,058			5,867			–	–
1DX (10′)	2,991			2,802			–	10,160

Source: Data from Computers & Industrial Engineering (2016), Polish Register of Shipping PRS (2012a, 15), Grzybowski et al. (1997, 75), ISO (2013, 3).

Considering their gross weight, it is possible to distinguish small containers weighing up to 2,500 kg, medium-sized containers weighing from 2,500 kg up to 10,000 kg and big containers that weigh more than 10,000 kg. ISO-type containers are classified in this system as large-size containers.

Apart from the standards for container dimensions and gross weight, the standards for the labelling of ISO containers have been also provided, as such containers must carry information referring to (Wiśnicki 2010, 22–23):

- identification of the owner of the container,[2] its identification number, the check digit, the type and its country code;
- specification referring to exploitation details, such as the maximum gross weight, marking indicating an open-top container, the height of over 2.6 m and information about limited stacking height; and
- additional information about meeting the requirements for railway transport (requirements of the International Union of Railways, UIC), adjustments for transporting hazardous cargo, perishable goods, liquid cargo (tank containers), CSC Safety Approval labels introduced by the requirements of the CSC Convention.

Hence, containers have become a part of a larger system. The container system has made it possible to implement efficient cargo shipping from consigners to consignees using maritime and land transport. One of the most significant features of the system understood in such a way is the unification aspect in the field of container construction (standardization).

2.2   THE CONSTRUCTION OF CONTAINERS

Various materials are used for constructing containers: steel, aluminum, plastic, plywood and/or wood. The basic element of a container is its frame, which supports the whole container structure. This is the part to which corner castings are welded (Figure 2.2). The frame sustains the heaviest strain, and it must support the gross weight of the container as well as the mass of other containers stacked on it (several containers are usually stacked one on another at storage yards and in cargo holds of vessels). Therefore, the particular parts of the frame come as the most robust elements of the entire shipping device. Corner castings are for lifting empty or full containers from various means of transport with the use of specialized cargo-handling equipment. Bottom corner castings are for stowing containers inside means of transport. The frame of a container is made of steel, aluminum or durable plastic. Other elements of a container are attached to its frame: the floor, sidewall panels, the end panel, the container door and the roof panel (Grzybowski at al. 1997, 78). The sidewall panels and the roof panel can be made of corrugated steel sheets, wood and metal laminates, plastic. The construction structure of sidewall panels and of the roof panel is presented in Figure 2.2.

Similarly to the frame structure, the floor of a container must be highly durable because it holds the weight of the entire cargo during shipping and storage. The floor consists of strong longitudinal beams and cross members covered with floor

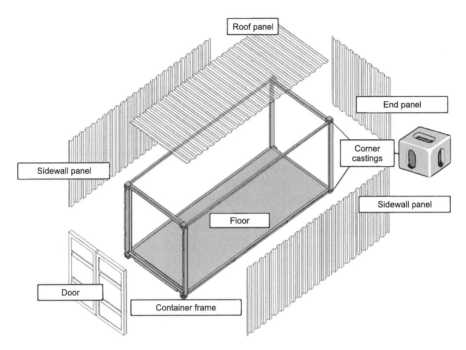

FIGURE 2.2 The construction of the roof and sidewalls of an ISO container.

Source: Adapted from: www.residentialshippingcontainerprimer.com/conceive%20it (accessed: 7th February 2021).

panels made of plywood or plastic. The floor also contains two forklift pockets and a gooseneck tunnel,[3] which are needed to handle containers around terminal yards. A container may have one or more tight-sealed two-leaf doors used for stuffing and stripping. Most frequently, the doors are located in the front (narrow) panel, but sometimes they are set in the sidewall panels, which makes the access to the cargo much easier and facilitates thorough inspections of a container. The door is secured against an accidental opening with a locking-bar mechanism with two or four locks, on which padlocks or seals can be put, as required during the customs seal procedures.

The previously mentioned corner castings must be located in accordance with the adopted standards, and manufacturers are obliged to provide special care while attaching those parts to containers. The precision of that work decides about the compatibility of a container with cargo handling facilities (e.g., dockside gantry cranes, self-propelled reach stackers, railway gantry cranes, etc.). Technical standards are provided by the ISO and classification associations that supervise maritime containers.

The most important for container producers is to keep the precise distance between the opening centers in corner castings as they are different for each type of container. For example, the openings along the longer side of a 40ft container should be located at the distance of 11,985 mm, and along its shorter side, they should be located at the distance of 2,259 mm. The information on the materials and construction specifications are provided in detailed regulations for container construction, which are published by classification associations (in Poland, the supervision is provided by the Polish Register of Shipping).[4]

Depending on the material used for the construction of containers, their total weight can differ (Table 2.3). The lower its weight, the higher loading capacity of a container—but also the higher cost of its production. The most popular material is steel, considering its availability and low price. Insulated and reefer containers are often made of materials characterized by low heat conductivity. They make it possible for operators to maintain the required temperature inside containers.

During their exploitation, containers are exposed to sea water, rainwater and low and high temperature. Such conditions are favorable for the corrosion of their steel elements. Therefore, corrosion-resistant materials are used for the construction of containers (galvanized steel, metal and plastic laminates, reinforced plywood), which are then covered with the proper number of varnish coatings.

The structure of containers also depends on the scope of cargo groups that are going to be transported inside them or on one specific type of cargo. Despite the standardization of their dimensions and other important elements of their structure, containers are very different depending on the physical and chemical characteristics of the transported cargo.

Depending on the type of a container, it can be equipped with fittings necessary to secure the cargo inside. Most frequently, such elements can be found in general-purpose containers and various types of flat rack containers. The internal constructional elements of containers, including corner posts, the bottom frame and the floor, are usually equipped with special lashing points, which make it possible to use spring lashings to stabilize the cargo (Salomon 2017; Wiśnicki, ed. 2006, 86–89). Flat-rack containers and platform containers are equipped with robust fasteners of high durability as they are usually used for transporting heavy cargo. Stuffing and stowing general cargo inside containers requires the use of appropriate materials that can increase cargo safety on the way to its destination point. The most commonly used stowage materials are (Salomon 2017):

• Euro-pallets that isolate cargo from the container floor and provide free airflow;
• plywood panels for vertical and horizontal separation of particular lots of cargo;
• lashing belts for stabilization of cargo inside containers or on flat-rack and platform containers; and
• steel wire ropes, lashing chains and tensioners for proper lashing tension.

TABLE 2.3

Comparison of Container Weight, Depending on the Material Used for Their Construction

Container	Weight of containers (given in kg)			
	Plastic	Aluminum	Light steel	Steel
20'	1,565	1,760	2,300	2,400
40'	2,268	2,959	3,420	3,860

Source: Data from Grzybowski et al. (1997, 81).

The International Conference for Safe Containers (CSC) obligates each participant country to indicate an entity responsible for providing technical supervision over the production and exploitation of containers. The scope of the supervision provided by that entity includes the following (Polish Register of Shipping PRS 2012a, 10–12; Wiśnicki and Kotowska 2009, 4–6):

1. analysis and approval of technical specifications
2. supervision over container production
3. testing
4. marking and standardization
5. issuance of documentation
6. approval of plants and testing centers

Before the production of particular containers or their prototypes—which are then going to be implemented into series production—is started, it is necessary to deliver technical specifications to the appointed institution. The submitted documents should include specifications referring to construction, parameters, materials to be used for construction, methods of construction, technology of assembly, finishing and paint coating. Additionally, the manufacturer is obligated to provide a schedule of tests, after which containers are approved for use. The procedures of approving containers supplied by a particular manufacturer require submitting complete technical specifications and testing performed in the presence of an inspector who represents the appointed institution. If a container (a container prototype) meets the constructional requirements confirmed by testing, the institution issues a Certificate of Approval for a Container Type Construction (which is identical with container certification stating that it is safe and it meets the requirements compliant with the CSC Safety Approval certificate). A company that wishes to start series production of containers has to be approved by the supervising institution. It also means that the entire system of quality control needs to be approved as well. The required documentation includes a complete specification of container manufacturing processes and an organizational structure of the company, a system of quality maintenance (a quality manual), a system of quality control, manufacturing processes, technical supervision, logistics of production, employees' qualifications and individual manufacturing documentation of each container. Upon any request of a supervising institution, the manufacturer should provide the complete documentation of containers that have been produced and information about all changes and improvements that have been made to the container structure.

Based on the standards provided by the International Convention for Safe Containers, containers in operation are subject to the supervision of an authorized institution (in Poland the supervising institution is the Polish Register of Shipping). The scope of the supervision over containers in operation includes (Polish Register of Shipping PRS 2012b, 6):

- inspections and supervision over repairs and tests;
- approval of schedules for periodic and approved continuous examination programs for containers;
- authorization of plants for container repairs and testing centers;[5]

- periodic inspections of container owners and lessors, institutions, authorized plants for container repairs in terms of the proper course of container
  inspections and repair; and
- issuance of relevant documentation.

Containers must undergo their first periodic examination after five years of being in operation. The subsequent periodic examination takes place after thirty months from the date of the last examination. Approved continuous examination refers to the following situations (Polish Register of Shipping PRS 2012b, 7–8):

- A repair or a renovation of a container has been performed, or a container
  has been returned from the lease and is subject to the full inspection.
- An inspection for detecting damage that should be repaired is treated as
  routine examination.

Standardization also includes the principles for supervising the entire manufacturing process, quality control and individual technical specifications of each container. The stock of containers operated by maritime and land carriers and leased by logistics operators, importers or manufacturing companies undergoes inspections that are to ensure the proper technical condition and safety of containers during transportation. In the pragmatics of shipping processes, the question of selecting suitable types of containers for specific cargo becomes highly significant (which results from the assumed systematics).

2.3   TYPES OF CONTAINERS (TAXONOMY OF CONTAINERS)

The development of the container system has resulted in the implementation of various types of containers for transporting not only general cargo but also dry and liquid bulk cargo, technical gases and liquefied gases. The International Standardization Organization has introduced the following systematics for the basic types of containers (ISO 1496–3 2019):

- general cargo containers for general purposes
- thermal containers
- tank containers for liquids, gases and pressurized dry bulk cargo
- non-pressurized containers for dry bulk cargo
- platform and platform-based containers

The taxonomy of containers may consider their suitability for transporting all types of cargo, or general-purpose containers, or for transporting only some particular types of cargo, or special-purpose containers (Grzybowski et al. 1997, 86; Nierzwicki et al. 1997, 20). The diversification of containers also results from their various structure, which is adjusted to their functions or to the scope and ways in which they are used.

General-purpose containers[6] are characterized by a rigid structure based on a steel frame (which can be also made of any other robust material) to which the sidewall panels, the roof and the door are welded. A container of this type usually takes the form of a rectangular cuboid of different length and height, depending on its standard dimensions (1A, 1AA, 1AAA, etc.). General-purpose containers may be equipped with an open-top roof or a hardtop roof to facilitate crane operation during loading and unloading. Depending on their construction, the container doors can be located in the front wall at one end of a container, at both ends of a container or in its longer sidewall panels. Another constructional solution makes it possible to open the whole sidewall panels. The specific character of general-purpose container construction provides tight, insulated space where cargo is secured against the impact of external conditions (weather) and any damage that might be caused by tampering with collective or unit packaging.[7] The versatile character of this container type offers the possibility of loading various general cargo packed in pallet units (Sitko et al. 1970, 10),[8] individual pieces of cargo and homogenous general cargo (drums, canisters, sacks, cases, packets, bales). Versatile devices designed for shipping come as a large group, in which it is possible to distinguish other types of containers: ventilated, with openwork sidewall panels, half-height (half of the height of a standard container) (Grzybowski et al. 1997, 86).

There are also thermal containers used for transporting cargo. They allow operators to maintain the required temperature inside containers. This group of containers includes temperature-controlled containers for maintaining permanent temperatures inside, where the cargo can be cooled or heated. The structure of such containers is based on a supporting frame and sidewalls panels made of materials providing efficient insulation of the container interior from the external environment. Materials used for the production of thermal containers are usually various types of plastic used for making double laminated sidewall panels and special insulating dense foams. It is possible to distinguish the following types of thermal containers (ISO 1496–3 2019):

- insulated containers that function in a way similar to a thermos and do not have any heating or cooling supply sources
- refrigerated containers with no power supply, where ice, dry ice and cooling gas are used as the cooling factors
- refrigerated containers with built-in refrigeration units that require external power supply
- heating containers equipped with a heat source
- temperature-controlled containers equipped with air-conditioning and heating devices that allow operators to decrease or to increase the temperature inside containers

A complex structure and special equipment, including refrigeration units, heating and air-condition devices, result in the high cost of purchasing and operating such containers. The use of thermal containers in transport makes it possible to implement cold chains that are based on maintaining a constant temperature for cargo, starting on its manufacturing date to the date when it is collected by its ultimate receiver. This is to preserve the freshness of the cargo, to prevent it from decaying and to provide short-term storage (short-term storage is a procedure referring to shipping— e.g., when cargo has to wait for loading on a ship or a truck). Thermal containers are used for transporting agriculture produce (plant produce [vegetable fat] and livestock produce [meat, fish and other products]), frozen goods (ready-made food), pharmaceuticals (medicines, vaccines) and chemical products (surface active agents).[9]

A tank container is made of two basic elements: a framework that meets the standards of external dimensions (e.g., 1A, 1B, etc.) and a tank for transporting the cargo. Additional pieces of equipment are fittings for filling and emptying the tank and other indispensable elements, such as safety valves, analogue and digital gauges, tank-filling level gauges, thermometers, etc. Depending on their purpose, tank containers can be equipped with cooled or heated insulated tanks. Such containers are often used for transporting hazardous cargo or cargo that requires specific conditions, such as maintaining particular temperature or pressure. Hence, it is possible to transport a lot of liquid cargo types of various physical and chemical characteristics—for example, liquefied propane-butane (proper pressure), liquefied methane (must be cooled down to $-162°C$) or liquefied sulfur (must be heated up to $140°C$).

Another group includes containers the construction of which is adjusted for transporting dry bulk cargo. Shipping bulk cargo requires its loading and unloading; therefore, a number of various technical solutions have been implemented to facilitate those processes. Most frequently, dry bulk containers are characterized with the following technical features (ISO 1496–3 2019):

- filling openings in the roof used for loading only and sidewall flaps for handling the cargo
- a special flexible sack into which the cargo is loaded through the roof door and then the cargo in the tightly sealed sack is unloaded through the door in the sidewall panel
- an open-top roof that makes it possible to load cargo, which is then unloaded through the front panel door

Platform containers (flat-rack containers, bolsters) come as a large group of devices designed for shipping, and they are used for transporting cargo that cannot be transported by any other devices because of its size. The main element in the construction of a platform container is its floor, which can be additionally equipped in permanent or folded front and end sidewall panels. The dimensions of platforms correspond to the length defined by the ISO standards, 40ft and 20ft respectively. The most popular platform containers are the following (ISO 1496–3 2019):

- platform containers that do not have any vertical elements protruding outside the outline of the platform
- flat-rack containers with front and end sidewall panels that are attached permanently or can be folded
- platform containers that come as a combination of platforms and upper elements, such as sidewall panels, posts and foldable stanchions to keep cargo in place

Platform containers are mainly used for transporting individual pieces of cargo characterized by a large size and heavy weight, such as vehicles, machinery, construction elements, rotor blades for wind power facilities, yacht hulls, boats, etc. Empty platforms can be stacked one on top of the other and transported in this way in order to save space for other containers.

In considering economic aspects, foldable containers have been introduced into operation. They contribute to a decrease in costs incurred by maritime line ship-owners and other carriers who participate in shipping processes. Containers of this type are transported in their two forms: as unfolded and loaded containers and as folded and empty ones. What is important is it is possible to transport four of six folded containers stacked in a bundle instead of one unfolded container.[10] The big-gest advantage of applying this solution is the possibility to save space that can be used for transporting loaded containers. Considering the fact that such containers can be folded, their construction is more complex, and as a result, their tare weight is increased (from 500 to 1,700 kg). Also, their loading capacity is limited.

The most popular containers used for shipping are the following (Marek 2017, 34–37):[11]

- *Six in One—SIO*—they are 20ft containers that can be stacked into bundles of six after being folded. An SIO container is made of six elements that can be folded or unfolded (2,000 containers of this type have been produced so far).
- *Four in One*—they are 40ft containers which can be stacked into bundles of four after being folded; they can be folded with the use of hinges installed in the front and end sidewall panels.
- *Fallpac*—they are 20ft containers; their characteristic feature is an open top that comes as a separate element which is put on a container after unfolding the sidewall panels.

All types of foldable containers are suitable for handling various groups of dry general cargo. A problem with their exploitation can be the tightness of their con-struction. Hence, such containers are operated in a limited range by some maritime carriers, and their further development depends on advantages that can be gained during their exploitation.

Another group of containers includes special-purpose containers that are usu-ally adjusted for transporting some particular cargo. In terms of their construction, special-purpose containers can come as a combination of solutions discussed previ-ously; however, they also meet the requirements set by the parameters of the cargo, safety regulations for shipping such cargo and protection of natural environment. Special-purpose containers are used for transporting:

- hazardous cargo, including flammable materials, caustic liquid or solid substances that have to meet all safety requirements in accordance with relevant regulations;
- explosives, where thick walls are required as a safety measure, along with the possibility of secure locking (difficult to open by unauthorized parties);
- general cargo in technology of combining maritime and road transport, with the use of Transflat-type containers to which a top extension with a chassis can be attached (Trans.info 2021; Tradecorp 2021);
- cargo that has to be separated and that can be transported or stored in Duo-Con-type containers, which provide space equal to two separate 10ft containers by dividing a 20ft container into half, or in Quadracom-type

containers that offer space equal to four separate 10ft containers (RSD Container Industry Specialist 2021);

• pallets that require containers wider by 4 in. (10.2 cm). Containers of this type are usually referred to as pallet-wide containers, and they offer the possibility of loading two rows of pallets next to each other and, as a result, the possibility of transporting a higher number of pallets (20 PW—15 Euro-pallets, 40 PW—30 Euro-pallets, 45 PW—24 Euro-pallets) (Shipping and Freight Resource 2021);

• passenger cars transported on special racks inside containers (development by Trans-Rak International) (Global Trade Magazine 2021).

Containers that are most often used in transport are presented in Table 2.4.

TABLE 2.4
The Main Types of Containers

Container name	Container type	Scheme
General-purpose container	1C, 1CC, 1CCC 1[a], 1AA, 1AAA 1EE, 1EEE	
Open-top container, *Hardtop container*	1C, 1CC, 1CCC 1[a], 1AA, 1AAA	
Dry bulk container	1C, 1CC	
Refrigerated container (for refrigerated cargo)	1C, 1A,1AA, 1EE	

TABLE 2.4
The Main Types of Containers

Insulated *container*	1C, 1A, 1AA	
Flat-rack container, *platform container* A flat version	1C, 1A	
Flat-rack container, *platform container* A version with permanent front and end sidewall panels	1C, 1A	
Tank container	1A, 1C	
Foldable container Fallpac type	1C	

Source: Data from Computers & Industrial Engineering (2016), Polish Register of Shipping PRS (2012a, 15), CMD-Construction (2021).

To transport light, spacious cargo of a height that does not allow operators to load it into a standard container, 40ft containers of extended height of 9ft 6in. (2,895 mm) are used. They are usually referred to as high cubes. The extended height increases the cubic capacity of a high cube container by approximately 10m^3.

Containers have come as a revolutionary change to the processes of shipping general cargo, and they have made it possible to provide the complete the mechanization of their handling operations. For over 60 years of pragmatics pertaining to the operation of the container transport system, it has been developing and affecting all geographical regions and larger groups of mainly bulk cargo.

In the nearest future, it is possible to expect further development of containerization along with designing and implementing new types of containers that have already become elements of a bigger structure—the container transport system.

NOTES

1. Corner castings provide openings in the top and bottom parts of a container that can be used by gantries and cranes to handle containers during loading and unloading operations (as described further in the subchapter on container construction).
2. The code of the owner must be unique. It is registered at the Bureau of International Containers (*Bureau International des Container—BIC*) *Bureau International des Container—BIC*) or at any affiliated registration organizations based in particular countries. Check digits make it possible to verify whether the code groups have been read in the proper way. No IT system that handles a particular container will accept any identification marking if there is a conflict between the codes and check digits.
3. A special device used for handling low-bed semitrailers and containers, which comes as a part of the equipment at port container terminals, ferry terminals and general cargo terminals.
4. See more in Polish Register of Shipping PRS (2012a).
5. All repairs should be subject to the internal quality control system, and they should be compliant with the IICL (Institute of International Lessors) standards. The materials and spare parts should meet the requirements of maritime ship owners.
6. In accordance with the ISO standard 1496, general-purpose containers are also referred to as dry containers of general purpose; cf. Wiśnicki, ed. (2006, 71).
7. Packaging protects cargo against external conditions during shipping and storage processes. Packaging is not considered as a transport device, and it is of a disposable character. See more in Nierzwicki et al. (1997).
8. A pallet unit is a pallet with the cargo it carries. Pallet units can be handled with forklifts.
9. An example of a cold chain can be shipping vaccines against Covid-19, some of which require cold storage and a cold logistic chain in the temperature of −70°.
10. A bundle is formed by several foldable containers that are stacked one on top of the other and secured with a special security bolt for safe shipping. A bundle usually takes as much space as a typical 20ft or 40ft container.
11. See more in CMD-Construction (2021), Ramphul, Ramesh and Jaunky (2017, 36–38).

3 Container Transport System and Multimodal Transport Systems

3.1 THE CONTAINER TRANSPORT SYSTEM

The Container Transport System (CTS) is a specific organization of transporting containerized cargo that has been undergoing a process of optimization with the use of adequate technical elements, technologies, procedures and principles of management (Sitko, ed. 1974, 11). A sine qua non condition for the functioning of the CTS is the use of ISO-type containers that come as basic loading units to the entire system. Considering that aspect, the CTS is understood "a system of transporting and handling cargo packed into integrated (container) loading units" (Markusik 2013, 277). Under the CTS, various carriers operating in various modes of transport can cooperate in order to deliver cargo in ISO containers from its consigners to consignees (house to house/door to door) with the use of one or several means of transport. The CTS may be operated in the domestic as well as international ranges (in this context, also a global range). Basically, the system includes the following elements (Markusik 2013, 279–280):

- specialized technical means: ISO containers (in the full range of their type series), means of transport (rolling stock, such as railway platform cars for transporting containers, containers semitrailers, container vessels, inland waterway container barges), suprastructural means (cargo handling facilities, container tractors) and infrastructural means (land and port container terminals, depots, etc.)
- specialized technology for using technical means (e.g., stuffing cargo into containers at the consigner's premises and stripping containers at the consignee's premises, transporting containers by various means of transport, storing containers in transshipment between various handling operations)
- a dedicated organization and management system in the CTS with the use of IT/telematics technologies

The structure of the CTS is presented in Figure 3.1. Among the elements of technical equipment, ISO containers themselves come as particularly significant. They undergo constant improvement and modification that reflect the dynamics of changes taking place in the CTS. Apart from containers properly adjusted to transportation of general cargo, special-purpose containers have become more and more popular for transporting dry and liquid bulk cargo. The operation of the CTS requires proper

DOI: 10.1201/9781003330127-5

The location of port container terminals depends on the location of maritime ports at which they are located. They have currently formed a complex global network that allows operators to transport containers to almost any corner of the world. Land terminals are usually constructed in the vicinity of large urban agglomerations. A particular type of land terminal is a dry/rail port. Dry ports are in fact land terminals that are located in some distance to maritime ports and operated as their storage hinterland where the selection and coordination of containers take place, involving containers transported to the port and received directly from the port terminal that are to be taken to further hinterland destinations.

Indispensable technical equipment includes various types of cargo handling facilities that allow operators to implement loading, unloading and handling processes efficiently in the terminal area. The basic cargo handling facilities at port terminals are ship-to-shore (STS) cranes for handling cargo between the shore and the vessel. Similar facilities are also applied at land terminals. Additionally, each terminal has a number of devices for stacking containers and moving them around storage yards (various types of reach stackers, trucks, platforms, etc.).

Technologies for using the particular means of transport are related to their specificity, service life and technical wear. Cargo handling and auxiliary facilities and at container terminals are used for handling means of transport and providing handling operations.

The CTS requires an organizational system for the comprehensive and extensive cooperation of the entire cargo flow, from the consigner to the consignee. The development of a real transportation chain needs to be based on cooperation among all the participants of the container transport system. Managing cargo flows requires access to information that comes from various participants of the shipping process. Hence, the task of a coordinating entity is to synchronize the particular shipping stages and cargo handling operations. The efficiency of the CTS depends on the cooperation of all the entities that participate in processes pertaining to shipping, cargo handling, documentation handling, customs operations and other activities necessary for transporting cargo. The standard is to use various types of IT systems that provide numerous possibilities of storing and sharing information and supervising cargo flows. All container terminals operate with the use of modern terminal operating systems (TOSs) that are based on IT technologies that allow operators to identify particular containers, to verify cargo data, to submit shipments for customs clearance procedures, to inform forwarders about the current status of their cargo shipments (Markusik 2013, 279–280).

Under the CTS, in order to improve the possibilities of integrated management, various systems of container monitoring have been implemented along with systems dedicated to the monitoring of conditions and regimes of transport technologies (e.g., temperature, humidity), integrity of the entire set (a truck, a semitrailer and a container), presence of a driver in the vehicle cabin (and their identification according to the biometric documents), etc. An integrated electronic seal is a solution that has become more and more popular in that field. An example of a system for handling electronic seals is the RadioSecure (RS). Figure 3.3 presents the MASTER function (the main device) that is performed by an RS SLM seal put on a container/semitrailer door. Inside the loading space, there are two (auxiliary, reporting) SLAVE devices

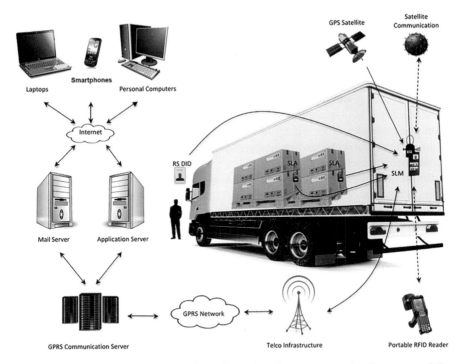

FIGURE 3.3 A scheme of transmitting information about events under the system of the RadioSecure electronic seal.

Legend:

RadioSecure is a conglomerate of MASTER managing devices, SLAVE reporting devices and an SCV internet platform used for comprehensive management of the devices and information on events they have signaled (alerts).

1. MASTER devices collect information from slave devices and communicate with the monitoring center:
   - RadioSecure SLM—a reusable seal for managing slave devices; it performs all GPS functions and it communicates with the monitoring center.
   - RadioSecure IVM—a device that is installed inside the truck cabin; it performs all GPS functions and it communicates with the monitoring center.

2. SLAVE devices communicate with managing devices and do not have any possibilities of independent communication with the monitoring center:
   - RadioSecure SLA—a reporting device; a reusable seal for reporting any tampering attempts to the MASTER device
   - RadioSecure SLE—a reporting device; a reusable seal for reporting any tampering attempts, humidity, temperature and light intensity to the MASTER device
   - RadioSecure SLL—a reporting device; a reusable seal for reporting any tampering attempts, humidity and temperature inside the loading space to the MASTER device
   - RadioSecure DID—a reporting device; a reusable seal for reporting any lack of integrity between the appointed driver and the truck cabin during the route to the MASTER device

Source: Reprinted with permission of Maciej Stawicki, CO GPSeal; www.plombyelektroniczne.pl/naczepa-slm--sla.html (accessed: 13th March 2021).

of the SLA-type installed on the pallets. If there were an additional SLE service (sensors) provided, the system would inform the managing entity about any changes of temperature, humidity, light and its own integrity. Then, in accordance with the scheme, the managing entity would report any detected anomalies to the monitoring center.

Applying integrated monitoring of containers and their cargo in the CTS improves the efficiency of management in the entire system. It also contributes to a decrease in insurance costs (considering the possibility of identifying each anomaly at each stage of the transportation process) and indirectly affects the level of security and safety in a container transport chain.

Entities that organize container transport processes may include forwarders, maritime carriers and port container terminals. Forwarders cooperate directly with cargo administrators, and under a forwarding order, they enter into shipping contracts and perform other work orders necessary to transport the cargo. Maritime carriers provide shipping services on the particular maritime lines; they can offer their services to cargo consigners, forwarders, logistics and manufacturing companies. If the consigner has some classic general cargo to transport, they can use services provided by port terminals, where general cargo is packed into container loading units (stuffing/ stripping) that can be transported further by sea.

Developing an efficient CTS requires involvement of numerous economic entities that have adequate technical means, capital and organizational possibilities. Usually, the CTS is of a global range and can provide transportation of cargo between various countries. The internationalization of maritime transport is possible because of the fact that the particular participants of transport chains come from various countries, and this fact does not interfere with the efficient implementation of shipping processes.

The development of the CTS is fostered by international regulations pertaining to the principles for shipping and shipping documents. At this point, it is possible to mention the implementation of multimodal bills of lading into the common use. They cover the entire shipping route of cargo transportation, from its consigner to its consignee (FIATA Multimodal Bill of Lading, MULTIDOC 95 and others) (Łopuski, ed. 1998, 489; Kubicki, Urbanyi-Popiołek and Miklińska 2002, 123).[1] The globalization of the world's economy and liberalization of international trade are also highly significant to the development of international transport systems.

ISO containers (as the key elements of the CTS) are commonly used not only in maritime transport but also in other modes of transport systems.

3.2   MODAL SYSTEMS OF CONTAINER TRANSPORT

The development of containerization in maritime transport has resulted in a necessity of providing comprehensive handling to entire transport chains, from consigners to consignees (under the CTS). The implementation of shipping processes requires the combination of various modes and means of transport. Nowadays, shipping processes (also of container units) are implemented practically in all the areas of the human natural environment, and because of that fact, shipping is characterized by separate immanent features assigned to the particular type of environment and specificity that affects methods, technologies, time and quality pertaining to the implementation of transport services (Miler 2016b, 34). In accordance with the most popular and commonly accepted classification that takes the criterion of vertical division of transport into consideration, transport can be divided into modes (Grzywacz and Burnewicz 1989, 46).

The criterion for that division is the character of the environment where means of transport are operated and which is determined by the type of a shipping route and the technique of moving along that route, which ultimately affects the type of means of transport.[2] At the first level of vertical classification, transport is divided into land, maritime and air (three basic types of environment). Considering the specificity of the particular environment types (considering available shipping routes and means of transport), further division allows us to distinguish six main modes of transport, namely road (automotive), railway (train), air, inland waterway, maritime and pipeline transport (Piskozub 1982, 19–20). Over the recent decades, as a result of civilization's progress and the conquest of the space, it has been also possible to observe development of space transport; however, it is not commonly considered as a typical mode of transport.

Road transport is characterized by the highest availability. Therefore, it is widely applied to implement most processes pertaining to general cargo shipping, including ISO containers. Practically unlimited possibilities of using public roads and toll motorways allow operators to transport cargo to any place that can be reached by a vehicle. This feature gives road transport a significant advantage over other modes of transport because its limitations are the smallest and it does not require any intermediary transport links (using other means of transport). Such possibilities result in the fact that road transport is used for providing feeder services, and it allows operators to reach the headquarters of any consigner and any consignee of the cargo. The significant feature of road transport is its high speed, which comes as an advantage when perishable goods, valuable goods requiring supervision (HVU—high value units) or dangerous, hazardous goods (DG or HAZMAT) are transported. The availability of road infrastructure and high speed of road transport enhance its flexible and versatile character. If a road network is available, road transport takes place also along international routes (Sitko, ed. 1974, 385).

Since the beginning of containerization, road transport has been applied as a link that provides hinterland container shipping (hinterland to ports and terminals). This means that it assumes an important function of delivering containers from consigners to port container terminals, from where the cargo is sent by sea later on. Analogical shipping processes are implemented between a port terminal, where the cargo is unloaded from a vessel and then needs to be delivered to its ultimate consignee.

In order to transport containers, trucks (tractor units) with semitrailers for one 40ft, one 45ft or two 20ft containers (2 TEU) (Mindura, ed. 2004, 95) are used. Trucks with frame chassis adjusted for transportation of one 20ft and a trailer for one 20ft container are used less often (Figure 3.4).

The main limitation of road transport is its low loading capacity as one truck can transport only two 20ft containers. The loading capacity of one road vehicle unit

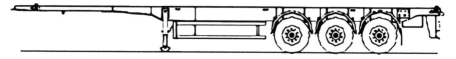

FIGURE 3.4   A container semitrailer for transporting containers in automotive transport.

Source: Adapted from www.paneltex.neostrada.pl/Polska_wersja/naczepa_kont.htm (accessed: 26th April 2021).

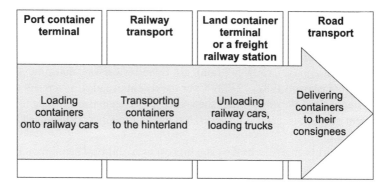

Port container terminal	Railway transport	Land container terminal or a freight railway station	Road transport
Loading containers onto railway cars	Transporting containers to the hinterland	Unloading railway cars, loading trucks	Delivering containers to their consignees

FIGURE 3.5   A scheme of a railroad container chain (section: a port terminal, a consignee of containers).

Source: The authors' own work.

consisting of a tractor and a semitrailer is highly limited when compared to means of transport applied in other modes of transport. Hence, it is necessary to involve a high number of vehicles that contribute to the intensity of road exploitation and to the pollution of the natural environment with exhaust emission.

A solution that is frequently applied in the pragmatics of land transport is handling railroad shipments, where it is possible to transport containers to remote hinterland and to deliver them directly to their consignees (Sitko, ed. 1974, 382). Usually, container trains are marshalled at a port container terminal, and containers are transported to a particular freight destination railway station or to a container terminal, where they are unloaded to a storage yard. The last stage of the shipping process involves loading containers onto trucks and delivering them to their consignee's headquarters or to any other destination indicated (Figure 3.5).

Railway transport comes as a very efficient system for transporting containerized cargo to medium and long distances. This mode of transport allows operators to achieve high carriage capacity because a container train that is 600m long takes up to 92 TEUs at a time (Kostrzewski, Nader and Kostrzewski 2018, 205; Table 3.1). Other advantages of railway transport include shipping speed and low environmental pollution when electric traction is used. Usually, medium- and long-distance routes are handled by freight trains with electric engines, whereas diesel engines are used for marshalling and maneuvering operations.

TABLE 3.1

Carriage Capacity of a 600m Container Train, Depending on the Type of Railway Cars (Applied in the EU Railway Systems)

Railway car type	The number of railway cars in a train	Length of a train (given in meters)	The number of 40ft containers in a train	The number of 20ft containers in a train
Rgmms	42	589.68	42	84
Sgmmnss	44	595.76	44	88
Sgrss	36	588.24	36	72
Sgs	30	597.00	30	90
Sgns(s)	30	592.20	30	90
Sggns(s)	23	596.62	46	92

Source: Data from Kostrzewski, Nader and Kostrzewski (2018, 205).

In Europe, the network of railway connections has been well developed; hence, there are not any problems with trains in transit traffic. In some Eastern European countries, different railway track gauges come as a limitation in comparison to Western countries. However, this technical obstacle has been already solved by changing railway car bogies or by using railway cars with variable gauge axles.

Basic limitations of railway transport include its lower availability, which is related to the necessity of developing some nodal infrastructure, such as freight stations or land railroad terminals where changing means of transport for containers could take place. The maintenance costs of linear railway infrastructure are very high as it consists of railway lines, electric traction systems and traffic safety systems. Considering replacement or development investments, high financial expenditures are needed, which may be several times higher than the costs of developing a road network.

Containers are transported by container trains marshalled on railway cars adjusted for transporting ISO containers of various types (Table 3.2). In fact, trains that leave port container terminals usually transport 40ft, 45ft and 20ft containers.

In railway transport, platform cars are often used with slots blocking containers during shipping and marshalling operations at railway stations and terminals (Figure 3.6).

Railway transport provides high transport capacity provided that railway infrastructure is properly developed and there is no congestion along the main railway trunk lines. In the European Union, it is considered that railways meet the requirements set for modern means of transport and provide a low emission of pollutants (if electric traction is taken into consideration).

Shipping containers from port terminals to hinterland can be implemented by inland waterway transport. The availability of those shipping services depends on the network density of natural and man-made inland waterways and river ports

TABLE 3.2

Types of Railway Cars and Their Container Transporting Capacity

Railway car types	Length of railway cars (given in ft.)	Carriage capacity for 20ft containers	Carriage capacity for 40ft containers
Rgmms	40ft	2 × C (2 TEU)	1 × A (2 TEU)
Sgmmnss	40ft	2 × C (2 TEU)	1 × A (2 TEU)
Sgrss	45ft	2 × C (2 TEU)	1 × A (2 TEU)
Sgs	60ft	3 × C (3 TEU)	1 × A (2 TEU)
Sgns(s)	60ft	3 × C (3 TEU)	1 × A (2 TEU)
Sggns(s)	80ft	4 × C (4 TEU)	2 × A (4 TEU)

Source: Data from Kostrzewski, Nader and Kostrzewski (2018, 203).

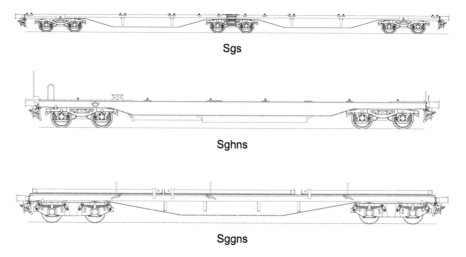

Sgs

Sghns

Sggns

FIGURE 3.6 Railway platform cars adjusted for transporting containers (top down: Sgs, Sgns, Sggns).

Source: Adapted from PKP Cargo Railway Cars Catalogue www.pkpcargo.com/media/1002903/pkp-cargo_katalog-wagonow_3008_19.pdf (accessed: 26th April 2021).

(Figure 3.7). The linear infrastructure should ensure safe navigation, which is related to proper depth and fairway marking (Kulczyk and Winter 2003, 62). Maintaining infrastructure in order to meet proper quality standards requires high investment in the channelization of rivers and construction of artificial channels (Kulczyk and Winter 2003, 40).

The main advantages of inland waterway transport include its carriage capacity and ecological nature. Depending on the type of barges or barge trains, container transport capacity can be from several dozen or several hundred TEUs. The size and draft of vessels must be adjusted to the particular water area where they are going to

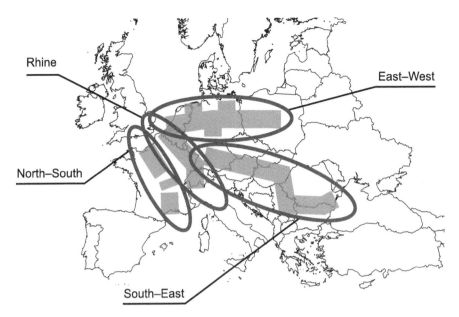

FIGURE 3.7   The main European inland waterways.

Source: Data from www.researchgate.net/figure/Overview-of-European-inland-waterways_fig1_273446772 (accessed: 21st July 2018).

be operated. Inland waterway transport is an option considered when port container terminals are located at river estuaries or artificial channels. This mode of transport may be beneficial, considering diversification of possibilities for transporting cargo to interior hinterland. The types of river barges adjusted for transporting containers by the European inland waterways are presented in Table 3.3.

Disadvantages of inland waterway transport include the low speed of shipping, a higher coefficient of the communication route and, in some countries, a ban imposed on the night operation of vessels with regard to navigation safety. Inland waterway transport may be successfully used for transporting containers with low delivery priority or empty containers. Shipping should be organized in the form of regular shipping lines between land and port terminals. Another limitation to inland waterway transport is the necessity for consigners and consignees to use feeder transport. This task is usually implemented by road transport and, less often, by railway transport (Kulczyk and Winter 2003, 40).

The modal transport systems are successfully applied in processes of shipping containers and most often used at the beginning and at the end of transport chains, where the critical links are relations with consigners and consignees (Figure 3.8). The efficient operation of the particular modes of transport is, in fact, aimed at obtaining the effect of complementarity of all the links and the efficient shipping of cargo to its consignee.

Based on the combination of the particular modes of transport into one chain from the consigner to the consignee of the cargo, the assumed method for shipping containers has proven to be efficient, and at present it comes as a typical transport solution.

TABLE 3.3

The Types of River Barges Adjusted for Transporting Containers (a Classification by the Criterion of Purpose and the EU Inland Waterway Classes)

Inland waterway class	Barge characteristics	Comparison to road transport
III	Container barge Length 63m, beam 7m, draft 2.5m Loading capacity 32 TEU	16 trucks (tractors with semitrailers)
Va	Medium container barge Length 110m, beam 11.40m, draft 3.0m Loading capacity 200 TEU	100 trucks
Vb	Large container barge Length 135m, beam 17m, draft 3.5m Loading capacity 500 TEU	250 trucks

Source: Data from www.bureauvoorlichtingbinnenvaart.nl/inland-navigation-promotion/basic-knowledge/waterways (accessed: 11th March 2018).

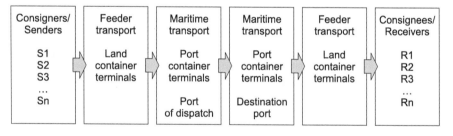

FIGURE 3.8   A scheme of a container transport chain.

Source: The authors' own work.

ISO containers (as the key elements of the CTS) are widely applied not only in modal but also in multimodal/intermodal forms of transport: multimodal transport, intermodal transport and combined transport.

3.3   MULTIMODAL (INTERMODAL) CONTAINER TRANSPORT SYSTEMS

In the pragmatics of processes pertaining to container transport, as a result of the modal division of transport, we often deal with containers transported with the use of stock characteristic for one mode of transport (e.g., "pure" road transport or maritime transport only). However, considering the specificity of contemporary foreign trade (internalization, globalization, longer distances, "door-to-door" imperative in transport services), more and more frequently, a necessity of combining modes can be observed for containerized cargo transport (ISO containers) under one container transport chain (among others in the CTS), where two or more modes of transport are used. The potential relations understood as possible intermodal transport combinations are presented in Figure 3.9.

In containerized cargo transport, the occurrence of intermodal operations significantly affects the level of complexity in cargo handling processes and handling containers in general (moving, rearranging, storing, stacking, checking, securing, stowing). It requires both: proper technologies and process organization. Figure 3.10 presents a model approach toward container handling in the intermodal system.

Aimed most frequently at the implementation of door-to-door shipping processes, such a complicated system of intermodal transport processes involving containerized units results in the fact that unification entities and processes must appear to integrate and to coordinate the entire transport system. In fact, they can be observed

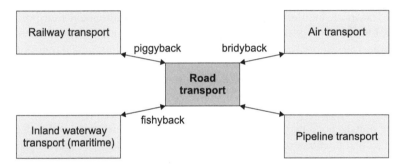

FIGURE 3.9   The potential of intermodal transport combinations.

Source: Data from Miler and Kuriata (2019, 114).

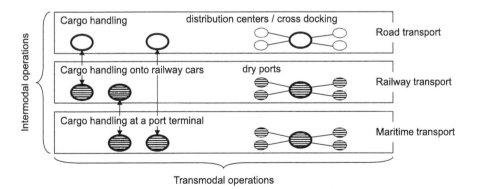

FIGURE 3.10   Container handling processes in the intermodal system (maritime to railway to road).

Source: Adapted from Rodrigue, J.-P. (2017).

at all the integration levels (the system of physical handling and operations involving loaded containers, the transactional-commercial system, the management system, the information and monitoring system, the system of liability for the cargo and documentation handling).

The previously mentioned factors, which determine the multimodal/intermodal transport forms, make it possible to distinguish multimodal transport, intermodal transport and combined transport.

MULTIMODAL TRANSPORT

The development of international container transport has underlain the beginning of multimodal transport, which is also referred to as intermodal transport, especially in maritime shipping. The term can be found in bills of lading issued by shipowners: multimodal bills of lading and FIATA multimodal bills of lading. It can be also found in documents published by the International Chamber of Commerce in Paris on the unified common practice pertaining to documentary letters of credit (multimodal transport, multimodal transport operators and multimodal transport documents). One of the very first definitions of multimodal transport describes this type of transport as shipping cargo in one and the same unit by intermodal transport, with the subsequent use of several modes of transport and without handling the same cargo in transshipment when the means of transport are changed. A unit in intermodal transport can be, for example, a container, a swap body, a road vehicle, a railway vehicle or a vessel Eurostat ITF 2021). The backhauling of empty containers or other loading units is not a part of intermodal transport. A disadvantage of this definition is the fact that it refers exclusively to technical and technological aspects of shipping processes implemented with the use of loading units. The definition omits organizational (multimodal transport operators) and legal (taking goods under the custody of the operator who is responsible for the transported cargo) aspects that are strictly related to multimodal transport (Neider 2012, 111).

Therefore, a more complete definition of multimodal transport defines it as a transport system applied to implement cargo shipping processes with the use of at least two various modes of transport under one contract for multimodal transport that covers the entire shipping route. The carrier is responsible to the consigner for the shipping process from the moment the carrier takes the cargo from its consigner to the moment when the cargo is passed to its consignee (Kubicki, Urbanyi-Popiołek and Miklińska 2002, 97–98; INFOCar.net 2021).

In multimodal transport, the key role is performed by multimodal transport operators who are carriers (shipowners or road transport carriers) or forwarders. Operators perform their duties from the moment of taking a container or any other loading unit under their custody to the moment when cargo reaches its destination. In practice, some other solutions can be found, in which operators, apart from tasks included into the complete transport cycle, provide additional services of delivering, stripping and shipping empty containers to a particular depot. Operators may enter into agreements with other carriers in order to extend and improve services they offer. As it does not release operators from their responsibility for safe and proper cargo shipping, they must coordinate and monitor work of their subcontractors (carriers).

Among multimodal transport operators, the following entities can be distinguished (Kubicki, Urbanyi-Popiołek and Miklińska 2002, 109–110):

- direct operators who handle shipping with their own means of transport
- indirect operators who handle shipping in cooperation with subcontractors

Multimodal transport allows parties to reduce transport costs and to shorten the time of delivering the cargo to its ultimate consignee. In order to achieve these aims, it is necessary to standardize loading units, means of transport and cargo handling facilities.

INTERMODAL TRANSPORT

A definition of intermodal transport is similar to the definition of multimodal transport. However, intermodal transport is based on the concept involving one operator (one carrier), one contract and one responsibility of the carrier under that contract and one price for the entire shipping process (Neider 2012, 111–112). The essence of this type of transport is well caught by the following definition: "intermodal transport means shipping cargo in loading units with the use of means of transport applied in at least two modes of transport, based on unified provisions stated in an intermodal transport contract which is entered by a customer and an intermodal transport operator" (Neider 2012, 111–112). In accordance with the previous definition, intermodal transport consists of four basic elements (Neider 2012, 112):

- using a loading unit for shipping means that the cargo is handled and stored together with that unit
- using means of transport applied in at least two modes of transport
- using only one contract for transporting the cargo along the entire shipping route
- employing only one contractor who is responsible for the organization and course of the whole shipping process and who uses one shipping document that covers the entire shipping route

Hence, in intermodal transport, cargo is transported in the same loading units, and it cannot be handled by any other loading units during the change of the mode of transport. Multimodality is based on the existence of alternative modes and means of transport along the same shipping route (or along a transport corridor), whereas intermodality is based on the use of several modes of transport under one integrated transport chain along the same shipping route (Wronka 2013, 25). The development of intermodal transport has been supported by the European Union as it contributes to a decrease in energy consumption and exhaust emission in transport and also to a decrease in road traffic congestions. Therefore, it is more ecological than road transport alone. An important factor of that development is the existence of infrastructure providing possibilities to handle loading units—e.g., containers from one means of transport onto another.

Intermodal transport involves the integration of shipping processes in three areas (Marciniak-Neider and Neider, eds. 2014, 467–471):

- Technical and technological—it involves adjusting means of transport and cargo handling facilities at land and maritime terminals to the handling of standardized loading units, such as containers, swap bodies and truck semitrailers.
- Organizational integration—it is manifested by the existence of intermodal carriers who can help their customers out of all the responsibilities related to the implementation of the shipping process.
- Legal integration—it means that there is one contract entered by an intermodal transport operator and the ordering party and there is one shipping document applied to cover the entire shipping route. If the operator relies on services provided by transport companies, the operator must enter into respective contracts with them; however, the operator is still solely responsible to the ordering party for the implementation of the shipping along the entire house-to-house route. The operator sets a total price for the organization and implementation of the entire shipping process to the ordering party, who can easily compare offers presented by various operators in terms of transport costs.

In the field of intermodal transport, it is possible to distinguish several ways of transporting cargo in loading units, which differ in their technical and organizational aspects (Kubicki, Urbanyi-Popiołek and Miklińska 2002, 95):

- transporting containers by land and sea, with the use of vehicles and/or railway and vessels
- transporting cargo by land and by ferries
- transporting cargo by railway and road

COMBINED TRANSPORT

In accordance to the Economic Commission for Europe and Eurostat (EU), combined transport is "intermodal transport of goods where the major part of the journey is by rail, inland waterway or sea and any initial or final leg carried out by road is as short as possible" (Eurostat ITF 2021). According to the Directive EU 92/106/WE, the distance should be less than 100 km for railway-road transport and 150 km for road-inland waterway combined transport or for road-maritime transport. Combined transport is defined in a similar way in the Act of 6th September 2001 on road transport, namely as "shipping cargo by trucks, trailers, tractors with semitrailers or semitrailers without tractors, swapbodies or 20ft containers or bigger containers which use or are transported by road along the Initial or final sections of the shipping route and they are transported by railway, inland waterways or sea along other sections of transportation, however the maritime section is longer than 100 km, measured as the crow flies" (Journal of Laws 2001 no. 125 item 1371). In combined transport, usually only one loading unit is used along the entire shipping route. At first, it is transported by a vehicle and then by a railway car or a river barge, and the last section of the route is covered by a vehicle again. The particular sections of the shipping process are usually characterized by their separateness and independence. It means that the particular carriers are responsible for the particular sections, and there is not any single entity responsible for the shipping along the entire route. Hence, the ordering party who contracts the shipping service needs to enter into a separate contract with each carrier—the total number of contracts corresponds to the number of means of transport applied in various modes of transport to ship the cargo along the entire route (Neider and Marciniak-Neider 1997, 18–19).

The development of combined transport is based on its ecological aspects. The heaviest burden is imposed on the natural environment by road transport, whereas the most ecologically friendly are railway and inland waterway modes of transport. Considering the fact that in numerous European countries, navigation along river networks is limited, inland waterway transport is of minor significance there. Hence, in European countries, combined transport is implemented as railway-road transport, where only short sections of shipping routes are covered by vehicles and then cargo is transported by railway. A specific form of combined transport is piggyback transport, in which trucks or tractors with semitrailers loaded with cargo are transported by railway. As a result, the same unit is transported by means of transport applied in two various modes of transport, where one (passive) means of transport is transported by another one (active).

Combined transport allows the advantages of railway transport (pro-ecological nature, relatively high loading capacity of railway cars, relatively cheap shipping for long distances, reliability of railway connections) to merge with the advantages of road transport (a possibility of delivering cargo in the house-to-house system, flexibility pertaining to the choice of shipping routes, high availability, various loading capacities of trucks). Nevertheless, combined railway-road transport is not without disadvantages, for example (Marciniak-Neider and Neider, eds. 2014, 479–481):

- higher transport costs when compared to road transport alone
- longer delivery times in the house-to-house system when compared to road transport alone because of the fact that it is necessary to stop the shipping processes at the initial and destination stations
- low punctuality of freight trains transporting loading units
- a small number of shuttle trains between land and sea container terminals
- a small number of land terminals adjusted to handle combined transport

Apart from technological and organizational regulations, the system of shipping containerized cargo with the use of several modes of transport is subject to systemic regulations in the formal and legal fields.

3.4 SYSTEMIC LEGAL REGULATIONS IN DOMESTIC AND INTERNATIONAL CONTAINER TURNOVER

3.4.1 INTERNATIONAL CONVENTION FOR SAFE CONTAINERS (CSC)

Using containers for shipping and in storage management poses three types of risk, the results of which might involve not only considerable material loss but also loss of human health or even life (Książkiewicz and Mierkiewicz 2014, 156–161):

- threats related with containers themselves as specific technical devices
- threats generated by cargo loaded in containers and erroneous cargo documentation (inadequate to the type of cargo and methods of securing such cargo for shipping)
- threats posed by containerized cargo movement (inside containers and containers themselves loaded onto vessels) resulting from erroneous stowage

In order to minimize the probability of these threats, the processes of stuffing, transporting, handling and storing containers are regulated by respective legal regulations, both at the domestic and international levels. Technical requirements that must be met by containers in operation are stated in the *International Convention for Safe Containers—CSC*. It was presented on 2nd December 1972 by International Maritime Organization and the United Nations. The document contains two annexes referring to technical aspects, namely (*Journal of Laws* 1972):

- Annex 1. Regulations for testing, inspection, approval and maintenance of containers; and
- Annex 2. Structural safety requirements and tests.

The convention refers to new and existing containers used in international transport, except for air containers. The document defines a container as a device that is durable and robust enough to be also reusable. Containers have to be equipped with corner castings for container handling, stacking and lashing to secure them for shipping by various means of transport. Starting as early as at the stage of construction, containers must be approved (certified) by an authorized institution.[3] Then they must undergo inspections—the first inspection takes place within five years from the container construction date. The subsequent inspections are scheduled every 30 months. The scope of the inspection includes (*Journal of Laws* 1984.24.118, Art. IV):

- visual examination to check the technical condition of a container;
- checking the right size and dimensions of a container, if there is a suspicion that they have ceased to be compliant with the respective requirements;
- testing for tightness against bad weather conditions, if there is a suspicion that a container has lost its tightness;
- checking the lock bars, doors, openings and any other elements of a container; and
- checking the container marking.

Each container that has been approved for operation must have a permanent, noncorrosive metal safety-approval plate attached to its wall. It is a rectangular plate, the dimensions of which are 200mm × 100 mm with CSC SAFETY APPROVAL inscription on it. The plate should present at least the following data (IMO CSC 1972, 18):

- an abbreviation of the approving country and approval certification data
- the date of container manufacturing
- the identification number of the container
- the maximum gross weight
- the permissible stacking weight
- the load value in the transverse racking test
- the date of the first inspection for new containers and the dates of the subsequent inspections during the container exploitation

The convention regulations are significant to containerized cargo transport because they provide comprehensive principles for approving containers into operation. Containers that meet the defined constructional and technological standards are the only ones approved for operation. The confirmation of meeting such parameters is a certificate in the form of a safety-approval plate. The ACEP (Approved Continuous Examination Programme) abbreviation means that the container is under constant supervision of a classification institution. The PES (Periodic Examination Scheme) means that the container undergoes periodic inspections (and that the period between inspections is defined based on the age of the container, the ways of its operation and current damages/failures).

Threats related to containerized cargo transport have been considerably reduced by the methods of stuffing or packing cargo into containers. In 2014, an updated version of *IMO/ILO/UN ECE Guidelines for Packing of Cargo Transport—CTU* (United Nations Economic Commission for Europe 2021) was adopted. Despite the fact that the document does not have legal validity, it is commonly applied in container transport as a set of good practice methods in the field of stuffing/packing containers. In the document, it is emphasized that the erroneous stowage of the cargo in a container or any other loading unit may result in the loss of human health for people who are involved in container handling. High material loss may also occur as a result of cargo damage inside a container. The most important recommendations included in the CTU Code are the following (United Nations Economic Commission for Europe 2021):

- During the operations related to stuffing containers, bad weather conditions and rough sea conditions should be taken into consideration for containerized cargo transported by sea.
- Special care should be taken during operations that involve container handling at port terminals where they are exposed to high acceleration and they are tilted and lifted by gantry cranes.
- Before stuffing, containers should be inspected in terms of their external and internal conditions: they should be clean, dry and free from any odors and remains of the cargo previously transported inside. The conditions of sidewall panels, floor doors and gaskets should be carefully inspected.

- The cargo inside the container should be stowed in accordance with the stowage plan.
- It should not be assumed that heavy cargo will not move inside the container; therefore, all pieces of the cargo must be securely lashed.
- The weight of the cargo must be evenly distributed on the whole surface of the container floor, and it cannot exceed the permissible loading capacity of the loading unit.
- Heavy cargo cannot be put onto light cargo; when stacking the cargo, the durability of the cargo packaging should be taken into consideration, and the particular cargo tiers should be separated with stowage dunnage materials.
- In order to avoid cargo damage done by moisture, wet cargo cannot be packed with cargo sensitive to moisture. Wet dunnage material, pallets and wet packaging should not be used as well.
- It should be checked if all the locking bars and seals are secured and locked properly after the container has been closed.

3.4.2 LEGAL REGULATIONS FOR TRANSPORTING HAZARDOUS CARGO IN CONTAINERS WITH PARTICULAR CONSIDERATION OF THE INTERNATIONAL MARITIME DANGEROUS GOODS CODE (IMDG)

There are some separate legal regulations provided for transporting hazardous cargo, such as flammable products and materials, explosives, radioactive, caustic and toxic materials, etc. They are also transported as containerized cargo, most often by intermodal transport systems: road–maritime or railway–maritime.

Improper handling of hazardous cargo poses a serious threat to human life, property and natural environment. Therefore, shipping such cargo requires special care and knowledge of legal regulations for transporting hazardous cargo, which must be strictly followed during all the stages of shipping processes. In Europe, international transport of hazardous cargo is regulated by three legal acts:

- In road transport, the *European Agreement concerning the International Carriage of Dangerous Goods by Road* (ADR) is applied (ADR European Agreement, Vol. I and II).
- In railway transport, the *Regulations concerning the International Carriage of Dangerous Goods by Rail* (RID) is applied (Intergovernmental Organisation for International Carriage by Rail n.d.).
- In inland waterway transport, the *European Agreement Concerning the International Carriage of Dangerous Goods by Inland Waterways* is respectively applied.

The previously mentioned regulations are supplemented by the *International Maritime Dangerous Goods Code* (IMDG) (IMDG Code 2018) applied in maritime transport. The document provides a classification of dangerous goods. It also provides regulations for construction of receptacles and packaging of dangerous cargo, methods of their marking and describing, regulations for shipping dangerous goods by sea (how to provide proper stowage, separation, lashing and inspections of the cargo), a list of all hazardous goods that can be transported by sea and a list of procedures related to fires and spills of dangerous substances, procedures related to reporting accidents involving hazardous cargo and guidelines for the stowage of hazardous goods inside containers.

The classification of dangerous goods in accordance with the IMDG Code is presented in Table 3.4.

Alternations and amendments to the SOLAS Convention made in May 2002 by the IMO resulted in the fact that the IMDG Code became a mandatory document, and since 1st January 2004, it has become applicable for shipping hazardous goods, their packing, marking, stowage in vessels, handling at the premises of ports/terminals and, in particular, storage and separation of noncompatible hazardous goods submitted for shipping.

The IMDG Code defines the conditions that must be met to minimize any risk posed by such cargo. These conditions include, among others, the following:

- applying the classification of dangerous goods in accordance with the UN recommendations
- selecting packaging compliant with the instructions referring to container stuffing
- proper container marking
- preparing all the required documents: a declaration of dangerous goods, a container packing certificate, a safety data sheet for dangerous goods
- providing training to people who work in the field of dangerous cargo shipping

---

TABLE 3.4
The Classification of Dangerous Goods in Accordance with the IMDG Code

Class	Name	Notes
1	Explosives	Subdivision into 6 subclasses
2	Gases	corresponds to Class 2 of ADR/RID
3	Flammable liquid materials	corresponds to Class 3 ADR/RID
4	Dangerous solid materials	Subdivision to subclasses corresponding to Classes 4.1, 4.2, 4.3 ADR/RIDIN
5	Oxidizing substances	Subdivision to subclasses corresponding to Classes 5.1, 5.2 ADR/I/ADN
6	Toxic or infectious materials	Subdivision to subclasses corresponding to Classes 6.1, 6.2 AIRID/ADN
7	Radioactive materials	corresponds to Class IDR/RID/ADN
8	Corrosive materials	corresponds to ClaI8 ADR/RID/ADN
9	Miscellaneous marine pollutants	corresponds to Iss 9 ADR/RID/ADN

Source: Data from Euro Shipping (2021).

According to the current legal regulations, the following principles should be followed for packing hazardous substances into containers (Wiśnicki, ed. 2006, 145):

- Stowage and lashing of hazardous cargo must be supervised by a person who knows the respective regulations, who is familiar with potential risks and who knows the procedures to be followed in case of an accident.
- Damaged cargo must not be packed into containers.
- Hazardous cargo must not be packed into the same container with any non-compatible substances.
- Barrels with hazardous substances must be always stowed in a vertical position.
- Containers with hazardous cargo must be labelled and marked with warning signs and placards.
- In maritime transport, a person responsible for stowage of hazardous goods in a container must submit a container packing certificate to the carrier to confirm that the cargo has been duly stowed, lashed and secured.

In the pragmatics of handling hazardous containerized cargo in maritime transport, there is another important element of risk that can occur when containers with dangerous goods are handled at port container terminals. Each terminal is characterized by its limited capacity of handling and storing containers with hazardous cargo. The factors that limit the capacity of storing hazardous cargo at a container terminal are the following (Miler 2016b, 36):

- the area and location of the main infrastructural elements of a terminal
- other vessels operating in the water area of a terminal
- location and number of warehouses and storage yards
- location and number of places designated for storing hazardous cargo in the general area of a terminal
- separate places designated for utilization and neutralization of hazardous cargo in an emergency situation
- the amount and class of hazardous cargo (including noncompatible goods that must be transported and stored in a considerable distance from each other)
- the hazard category assigned to the cargo in storage

During the planning stage of the processes pertaining to loading hazardous materials, all the previously mentioned components must be taken into consideration. The most significant problem is the necessity of proper separation (during the operation, in the area of a terminal and also during transportation in a hold of a vessel) of dangerous goods from each other when it is necessary to do so, according to safety requirements. As a result, it is also necessary to provide separate monitoring for maritime traffic of vessels carrying hazardous cargo (DGMT—dangerous goods maritime traffic)—as one of the image layers, separate monitoring for logistics processes (cargo handling, storage, temporary storage) involving hazardous cargo (DGLM—dangerous goods location and maintenance) and separate monitoring for conditions in which hazardous cargo is transported by sea (separation, stowage, technical parameters [e.g., humidity or temperature]; DGSC—dangerous goods shipping conditions).

Shipping containerized cargo by sea (including dangerous goods defined by the IMDG) significantly determines the development of the global fleet of container vessels.

NOTES

1. Generally in multimodal transport, it is possible to deal with the following documents: FBL-*Negotiable FIATA Combined Transport Bill of Fading*, replaced in 1994 by *Negotiable FIATA Multimodal Transport Bill of Lading*, FIATA FCR—*Forwarder's Certificate of Receipt*, FIATA FCT—*Forwarder's Certificate of Transport*, AIR WAY BILL (*nonnegotiable*)—recommended to forwarders by FIATA, FWB—(*nonnegotiable*)—FIATA *Multimodal Transport* FIATA, FWR—FIATA *Warehouse Receipt*, FIATA, FIATA SDT—*Shippers Declaration for the Transport of Dangerous Goods*, FIATA SIC—*Shippers Intermodal Weight Certification*, FEI—FIATA *Forwarding Instruction*.
2. This division is technical and exploitative in nature; however, in specialist literature, it is most often referred to as the modal division.
3. In Poland, as an example, the authorized institution is the Polish Register of Shipping.

4 The Development of the Global Fleet of Container Vessels

4.1 TYPES OF CONTAINER VESSELS

Approximately 90% of goods in global trade are transported by sea. Divided by product, the structure of maritime trade is highly diversified; hence, shipowners have to use various types of vessels. If the operation of a vessel is to be profitable, the vessel has to be adjusted in terms of its type, size and operational characteristics, not only to the demand for maritime shipping but also to shipping routes, port conditions, competitive vessels and other factors. This also refers to container vessels, which are usually divided by the following criteria (Miotke-Dzięgiel 1996, 47–48):

- navigation range (determined by the size and capabilities of passing through infrastructural objects, such as canals and lock chambers, e.g., the Panama Canal—Panamax)
- container handling methods (vertical, horizontal, mixed)
- vessel construction, which determines whether the vessel can be used exclusively for shipping containers or for transporting non-containerized cargo as well (con-ro, ro-pax and others)

Generally, it should be assumed that the diversification of sizes and types of container vessels depends on the economy of scales, which can be clearly observed in the hub-and-spokes concept, where large terminals are constructed as hub ports to concentrate cargo and smaller feeder ports. Such a solution is applied for a number of reasons, most frequent of which are vessel operating costs, reduction of pollution caused by large vessels (mainly exhaust emission), limitations to the size of port infrastructure related to the depth of fairways, port canals and basins and limitations to technical capabilities required to handle large container vessels at smaller container terminals.

Based on the criterion of the navigation range, it is possible to distinguish ocean-going container vessels (large and fast vessels designed for container shipping only; these are the VLCS class vessels—Very Large Container Ships),[1] medium-size container vessels of the Panamax[2] class and smaller feeder vessels that transport containers from and to hub ports, under the transhipment procedures (vessels of lower shipping capacities and lower speed, usually of the feeder[3] class, with independent onboard cargo handling systems). This type of vessel is usually used for shipping containers under cabotage feeder services and along shipping routes to minor ports

DOI: 10.1201/9781003330127-6

in developing countries (mainly African countries), where large and medium-size container vessels cannot call because of their technical (maneuvering) limitations.

However, considering typical communication routes handled by container vessels in the world, the Panama Canal is the key determinant of the vessel size. Before its modernization, the Panama Canal was able to handle vessels of the following parameters—beam 32.31m; length 294.13m; maximal draft 12.04m—which were referred to as the Panamax-type vessels (hence, the post-Panamax type refers to vessels with the beam above 32.31m). The expansion and modernization of the Panama Canal (the modernization work was finished in June 2016) resulted in a change to that terminology. A new category of Panamax (Neopanamax) container vessels have appeared, and this term refers to ships of the following parameters—beam 49m; length 366m and draft15.2m—which are adjusted to the parameters of the modernized Panama Canal. The New Panamax-type vessels are adjusted for shipping containers arranged in 19 rows and their capacity is about 12,000 TEU. The evolution of container vessels is presented in Table 4.1.

Considering the criterion of container handling technologies at ports, it is possible to distinguish vessels with vertical, horizontal and mixed cargo handling systems. In container shipping, vessels with vertical cargo handling systems dominate. Their holds are divided with metal guide rails into cells, the dimensions of which correspond to 20ft or 40ft containers (bigger containers, such as 45ft, 48ft or 53ft, do not meet these dimension standards, and they have to be transported on the top deck of a container vessel).

Ro-ro ferry vessels and barge vessels are also adjusted to vertical cargo handling operations. Ro-ro vessels are equipped with several continuous decks, some of which can be suspended or folded, depending on the requirements. In order to move cargo

TABLE 4.1
The Size of Container Vessels (1956–2020)

Year of entry into service	Vessel type	Vessel length	Capacity in TEU
1956	Early container vessels	137–200 m	500–800
1980	Full container vessels	215 m	1,000–2,500
1985	Panamax	250 m	3,000–3,400
1988	Panamax Max	290 m	3,400–4,500
2000	Post Panamax	285 m	4,000–5,000
2006	Post Panamax Plus	300 m	6,000–8,000
2012	Post New Panamax	397 m	15,000
2013	Triple E	400 m	18,000
2014	New Panamax	366 m	12,500
2018–2020	Ultra Large Container Vessel (ULCV)	≥400 m	>20,000

Source: Data from Nowosielski (2017, 110).

inside such vessels, various systems of ramps, lifts and trucks are provided. Loaded with containers, trucks, railway cars, semitrailers and special platforms are rolled straight onto the cargo decks of a ro-ro vessel. Containers can be also loaded on the top deck with the onboard cargo handling facilities. Considering the previously mentioned classifications, it is also possible to provide a synthetic classification of vessel types that are related to ferry and container shipping,[4] namely:

- ro-ro (roll on/roll off)—adjusted for transporting vehicles and rolling stock cargo;[5]
- con-ro—a combination of a container vessel and a ro-ro vessel, where the "con" part of the loading space (most often the bow part with holds equipped with cell guides, as on a typical container vessel) is designed exclusively for transporting containers, whereas the other "ro" part (usually located at the stern, with holds where the loading space is divided with tween decks) is designed exclusively for transportation of rolling stock cargo;
- ro-lo—where some parts or the entire loading space are adjusted for transporting cargo in the ro-ro system (horizontally, rolling stock cargo) as well as in the lo-lo system (lift on/lift off in a vertical system for palletized and containerized cargo);
- sto-ro—where cargo is loaded inside with the use of wheeled loading trucks (so there is an element of handling cargo in the ro-ro system) and then stowed and stacked there (so the "sto" part comes from stowage)[6];
- ro-pax—which is a type of a passenger and vehicle ferry, characterized by an extended loading space (for ro-ro loading and unloading processes) and a lower number of passenger cabins ("pax" comes from passenger); and
- other cargo handling solutions that combine the previously mentioned technologies in various configurations (ro-ro, lo-lo and sto-ro).

After 2018, a considerably higher number of the ULCV class vessels appeared. Ten of the largest container vessels in the world are presented in Table 4.2 as of January 2020.

Ship constructors and shipyard industry have not said their final word on the maximal size of container vessels. Hudong-Zhonghua, a Chinese shipyard, is planning to construct a container vessel of the capacity of 26,000 TEU, length of 432m and beam of 63.3m. The latest report of McKinsey Consulting Group indicates that it is very likely for ULCV container vessels of the capacity of 50,000 TEU to appear by the year 2067.

Considering the structure of vessels and combination of cargo they transport, it is possible to distinguish full cellular container ships and semi-container ships. Full cellular container vessels dominate in contemporary container transport. They have neither tween decks nor any other construction elements that would obstruct moving containers around the ship or unloading them with the use of wharf cranes. Inside vessel cells, containers are secured against moving during transportation. This is done with the use of permanent or movable metal guide rails. Containers are also protected against water with light covers put on the top layer of containers. There are

TABLE 4.2

Ten of the Largest Container Vessels in the World (as of January 2020)

No.	Name	Class	Flag	Length (given in m.)	TEU
1	HMM Algeciras	Megamax-Klasse	Panama	400	23,964
2	MSC Gülsün	MSC Megamax-24	Panama	400	23,756
3	OOCL Hong Kong	G-Klasse	Hong Kong	400	21,413
4	COSCO ShippingUniverse	COSCO-20.000-TEU-Typ	Hong Kong	399.9	21,237
5	CMA CGM Antoine de Saint-Exupery	CMA CGM 20.600 TEU-Typ	France	400	20,766
6	MadridMærsk	Triple-E-Klasse	Denmark	400	20,568
7	EverGolden	Evergreen 20.000-TEU-Typ	Panama	400	20,388
8	MOL Triumph	20.000-TEU-Typ	Panama	400	20,000
9	MSC Jade	MSC Pegasus-Klasse	Liberia	398.45	19,439
10	MSC Maya	Olympic-Serie	Panama	395.40	19,424

Source: Data from Ingenieur.de (2021).

also systems for pumping out rainwater or seawater that might get into a vessel with high waves. Smaller container vessels of the capacity up to 3400 TEU can have their onboard cargo handling equipment, but larger ones usually have to use port/terminal cargo handling facilities.

Semi-container vessels are multipurpose ships that are adjusted for transporting both containers and general cargo. The holds and decks of a semi-container vessel are designed in such a way that 40%–80% of the total capacity of the vessel can be dedicated to transport containers and the rest to transport general cargo. Semi-container vessels provide shipping connections to ports based in developing countries, where it would be difficult to load the whole vessel with containers. Considering the possibility of the flexible adjustment of multipurpose vessels to shippers' requirements, they are highly popular in contemporary maritime trade. General cargo ships that can be also partially used for transporting containers are less common. They are equipped with widened hold openings, and it is possible to install guide rails for a particular sailing in order to ensure the stability of containers during their transportation. The operation of semi-container vessels is economically justified particularly along the shipping routes where bulk cargo (such as raw materials) is transported to a particular destination, and containerized general cargo is shipped back from that place.

When operated in a proper way (regular inspections, replacement of worn-out vessel elements, necessary repair work) a period of technical operation of a container vessel is up to 40 years. However, an actual operation period is shorter, and it often does not exceed 30 years[7] because of the fact that more modern vessels are

TABLE 4.3

Countries Where Container Vessels Were Constructed at the Beginning of 2020

Specification	South Korea	China	Japan	Europe	Asia (excluding South Korea, China and Japan)	America
Number of vessels	1,788	1,497	649	963	432	25
Capacity (given in TEU thousand)	12,330	4,545	2,417	2,142	1,449	63
TEU share in %	53.7	19.8	10.5	9.3	6.3	0.3

Source: Data from ISL Shipping Statistics and Market Review (2020, 29).

introduced into the market and the competition among shipowners has become very strong. There is also a market of secondhand container vessels, where vessels from various shipowners' bankruptcy estate are sold along with vessels from shipowners who have been modernizing their fleets, or they have been experiencing problems with overcapacity.

For many years, the world's largest producer of container vessels has been South Korea. At the beginning of 2020, 53.7% of the global container tonnage that had been in operation was built in Korean shipyards. The next positions were held by China with 19.8% and Japan with 10.5% (Table 4.3).

The size and structure of the global fleet of container vessels have been undergoing dynamic changes as well.

4.2 THE SIZE AND STRUCTURE OF THE GLOBAL FLEET OF CONTAINER VESSELS

Over the last four decades, the number of full cellular container vessels in the world as well as their shipping capacity given in DWT have been growing year by year. The growth rate of the container vessel fleet after 2010 became a little slower than the world's maritime fleet, and as a result, its percentage share in the tonnage of the world's entire freight fleet decreased from 14.9% in 2011 to 13.9% in 2020, as seen in Table 4.4 (ISL Shipping Statistics and Market Review 2020, 10).

In 2019 the global fleet of container vessels was enlarged by 157 new vessels of the capacity of DWT 11.2 million (by 23 vessels fewer than in 2018). Among those new vessels, there were 10 ships of the capacity higher than 23 000 TEU. At the same time, there were 89 container vessels of the capacity of 2.4 million DWT withdrawn from service (in 2018 there were 66 such vessels of the capacity of 1.6 million DWT). At the beginning of 2020, the number of orders for new container vessels was 364 ships (26 million DWT; 2.4 million TEU), and it was lower than in 2019 (428

TABLE 4.4

The Global Fleet of Full Cellular Container Vessels in the Years 1980–2020

Year	Number of vessels	DWT in million	TEU in million
1980	606	10.3	0.7
1985	970	17.3	0.9
1990	1,169	22.3	1.5
1995	1,771	38.9	3.8
2000	2,287	63.6	4.2
2001	2,406	69.1	4.7
2002	2,524	77.3	5.3
2003	2,640	83.3	5.9
2004	3,036	91.6	6.4
2005	3,206	100.2	7.1
2006	3,494	112.7	8.1
2007	3,904	128.3	9.4
2008	4,276	144.7	10.8
2009	4,628	161.9	12.1
2010	4,706	169.2	12.9
2011	4,882	183.7	14.1
2012	5,083	196.4	15.3
2013	5,123	206.7	16.2
2014	5,141	216.6	17.1
2015	5,132	228.0	18.2
2016	5,274	244.5	19.7
2017	5,183	245.9	20.0
2018	5,190	253.0	20.8
2019	5,301	265.7	22.0
2020	5,360	274.7	22.9

Source: Data from UNCTAD (1985, 21), UNCTAD (1990), UNCTAD (1995), UNCTAD (2001), UNCTAD (2003), UNCTAD (2006), ISL Shipping Statistics and Market Review (2020).

ships; 30.5 million DWT; 2.9 million TEU). At the beginning of 2020, the number of vessels of the capacity of 1,000 < 2,000 TEU was the highest in the global fleet of container vessels—there were 1,257 such ships (ISL Shipping Statistics and Market Review 2020, 10). However, the largest share in the total capacity of container vessels given in TEU was taken by vessels of the capacity > 14,000 TEU, which accounted for 18.7% of the total container tonnage (Table 4.5).

The largest fleet of container vessels belongs to Germany. At the beginning of 2020, Germany had 1,051 container vessels of the capacity of 3,774,000 TEU, which accounted for 16.4% of the global fleet of container vessels. The next positions were held by China, Denmark, Greece and Japan. Totally, these five countries accounted for 61.5% of the capacity of the world's fleet of container vessels. The majority of all container vessels in the world (79.9%) are operated under foreign flags. The share of foreign flags is lower than 50% only in three countries, namely Denmark (44.6%), Indonesia (9.4%) and Vietnam (9.3%), as seen in Table 4.6.

TABLE 4.5

The Structure of the Global Fleet of Container Vessels Considering the Vessel Size (2020)

Size/Capacity (TEU thousand)	Number of vessels	DWT thousand	TEU thousand
< 1,000	1,053	8,527	631
1,000 < 2,000	1,257	23,833	1,776
2,000 < 4,000	920	34,622	2,574
4,000 < 6,000	826	49,309	3,939
6,000 < 8,000	270	22,259	1,802
8,000 <10,000	479	50,673	4,228
10,000 < 14,000	307	41,583	3,714
> 14,000	248	43,853	4,294
Total	5,360	22,948	22,948

Source: Data from ISL Shipping Statistics and Market Review (2020, 10).

TABLE 4.6

20 Countries with the Largest Fleets of Container Vessels (Vessels above 1,000GT)—January 2020

No.	Country	Number of vessels	DWT 1000	TEU 1000	Foreign flag (in %)
1	Germany	1,051	46,374	3,774	84.4
2	China	737	42,737	3,677	79.4
3	Denmark	348	29,354	2,538	44.6
4	Greece	478	25,058	2,041	97.7
5	Japan	316	22,408	2,005	88.5
6	Italy	221	18,340	1,521	100.0
7	France	150	13,622	1,176	79.7
8	Taiwan	256	12,379	1,002	82.7
9	Canada	113	10,455	911	99.9
10	Great Britain	191	11,237	905	88.8
11	Singapore	240	10,233	831	65.0
12	South Korea	181	6,956	558	74.9
13	Norway	76	4,410	366	100.0
14	United States of America	82	3,218	246	71.9
15	Turkey	89	2,655	212	71.2
16	Indonesia	223	2,579	177	9.4
17	Israel	32	1,913	155	83.1
18	United Arab Emirates	68	1,684	127	96.1
19	Iran	25	1,123	86	b.d.
20	Belgium	23	932	74	76.9
	World	**5,322**	**273,229**	**22,815**	**79.7**

Source: Data from (ISL Shipping Statistics and Market Review 2020, 13).

TABLE 4.7

The Container Vessel Fleet Out of Operation During the Years 2014–2020

The size of vessels given in TEU	I.2014	I.2015	I.2016	I.2017	I.2018	I.2019	I.2020	V.2020
< 1,000	39	42	43	60	24	26	33	58
1,000 < 2,000	55	32	64	70	20	81	58	131
2,000 < 3,000	24	26	53	35	7	33	34	64
3,000 < 5,099	69	9	80	99	14	41	41	106
5,100 < 7,500	24	2	57	39	3	3	12	61
> 7,500	8	5	40	48	14	11	75	131
Number of vessels	219	116	337	351	82	195	253	551
Share in the total number of container vessels (in %)	4.6	2.4	6.4	6.7	1.8	3.7	4.7	10.3
TEU 1,000	693	240	1,349	1,406	301	561	1,405	2,723
Share in the total capacity of the fleet (in %)	4.1	1.3	6.8	7.1	1.4	2.6	6.1	11.6

Source: Data from ISL Shipping Statistics and Market Review (2020, 14).

Considering vessels of the deadweight tonnage equal or higher than 300 GT only, at the beginning of 2020, there were 1,650 freight ro-ro ships in the world. They accounted for the total DWT of 8 million. There were also 2,978 freight and passenger ro-ro vessels of the total DWT of 4.7 million (ISL Shipping Statistics and Market Review 2020, 36).

Since the global financial and economic crisis of 2008, it has been possible to observe the overcapacity of the container vessel fleet that exceeds the demand for containerized cargo shipping by sea in the market of maritime container transport services. As a result, a part of the container vessel fleet remains out of operation (Table 4.7).

Problems related to overcapacity have become clearly visible as a result of an economic crisis caused by the Covid-19 pandemic.

4.3   THE IMPACT OF THE COVID-19 PANDEMIC AND A WAR IN UKRAINE ON THE GLOBAL FLEET OF CONTAINER VESSELS AND THE PROSPECTS FOR FURTHER OPERATION OF THE FLEET

In April 2020, a lockdown was imposed on 4.2 billion people (54% of the global population) by a number of governments in order to prevent the pandemic from spreading. It negatively affected 90% of economic activities in the world and it was the most acute for the service sector. Over 100 countries closed their borders, and as a result, foreign tourism and international trade relations were dramatically limited (UNCTAD Stat. 2020c). The death of numerous employees, high incidence of coronavirus infections and necessity of staying at home for thousands of people who had to undergo the quarantine procedures or to take care of children who could not attend their schools and nursery schools reduced the supply of workforce at places where business activities had not been stopped. The lockdown almost entirely paralyzed the catering, hospitality, tourism, sport and entertainment sectors. The pandemic also disrupted industrial operations. A necessity of keeping a safe distance between employees at production plants forced some companies to alternate their production processes if possible or to cease their operation completely. As a result, a decrease in production was observed; for example, in April 2020, the global industrial production decreased by 15% in comparison to December 2019.

A decrease in production in China resulted in a decrease in export, and in this way, it contributed to a reduction in container handling at Chinese ports—in February 2020, it fell down by 18% in comparison to the previous month (Abel-Koch and Ullrich 2020, 1–2). China plays a key role in international supply chains; therefore, any stoppage or decrease in supplies of components from China indispensable for manufacturing final products in numerous countries seriously disturbs production processes in thousands of plants that rely on Chinese semifinished goods. Some difficulties with the supply of indispensable products have been also observed in companies where manufacturing is based on raw materials and components imported from other countries.

Furthermore, the coronavirus has also hit the economy severely through a decrease in demand for a number of consumption and investment goods because consumers have considerably limited their purchases. Some people have been forced to do so because they have been laid off from their jobs as a result of the crisis and the lack of funds; others because they have started to buy less because of pandemic restrictions. Some people have decided to postpone or to resign from various purchases and to save money "for a rainy day." A decrease in demand results in the fact that companies produce, sell and export less. Hence, the tax revenues of the state also become lower. A decrease in revenues or their total loss puts credit borrowers in a difficult position—companies and households are not able to pay their credit liabilities. Such a situation results in a deterioration in the quality of the credit portfolio; it forces banks to create earmarked reserves for difficult credits, and it may become a reason for a bank crisis. All those manifestations of worsening economic conditions, international trade, internal consumption and internal demand affect international maritime transport.

The Covid-19 pandemic has severely affected international maritime trade and, most of all, container shipping by sea, through a decrease in the global trade volumes and disturbances observed in operation of port terminals in the world. Considering maritime transport, the shipping of consumption and investment goods has suffered more than shipping bulk goods. It should be emphasized, however, that global trade suffered from weakening much earlier during the entire period, from the beginning of the financial and economic crisis of 2008 to 2020, mainly because of some stronger protectionism tendencies in global economy and the trade war between the USA and China. In 2008, the total of global export and import of goods in relation to the global GDP was 51%, and in 2019, it dropped down to 43%.[8] Breaking international supply chains during the Covid-19 pandemic only accelerated de-globalization processes observed in global economy. According to UNCTAD, during the first quarter of 2020, the global trade volumes were decreased by 5%, and during the second quarter of that year, they were estimated to be at the level of 27% (United Nations 2020).

The lower Ilobal trade volume indicates less containerized cargo, lower demand for containers and container vessels and a lower number of containers handled at ports and transported by sea. According to Clearkson Research, global container handling at seaports in March 2020 was 5% lower than it was in March 2019. A decrease in the global trade volume year to year in April 2020 was 3%, and in May it was 7%. In May 2020, it was possible to observe that the volumes were lower at ports in the Republic of South Africa (21.7%), Southern Europe (18.7%), eastern and western coasts of South America (respectively 16.3% and 13.8%) and in ports of Southeast Asia (13.3%). At the same time, the number of containers laid off from operation was increased because of the lower demand for shipping services and the lack of cargo—from 117 units of the capacity of 336,000 TEU in July 2019 up to 551 units (2,723,000 TEU) in May 2020. It means that 11.6% of the capacity available in the global fleet of container vessels was unused (ISL Shipping Statistics and Market Review 2020, 5–8).

During the first half of the year, the oversupply of container vessel tonnage resulted in a decrease in charter rates charged in the market of container shipping. Charter rates for large container vessels became the lowest. Considering container vessels of the capacity of 8,500 TEU, a daily charter rate dropped by 48%, from USD 29,600 to USD 15,400, from the beginning of 2020 to May 2020. Simultaneously, a daily charter rate for container vessels of the capacity of TEU 2000 was decreased by 24%—from USD 9,350 to USD 6,900 (ISL Shipping Statistics and Market Review 2020, 9). The situation in the market of container shipping started to change during the second half of 2020, when almost the entire market of maritime transport was put into disorder. Shipping containerized cargo was most affected along the routes between China and Europe. It appeared that shipowners withdrew container vessels' tonnage that exceeded the drop in the demand for shipping containerized cargo from the market. This situation was overlapped with serious problems resulting from the lack of empty containers in Asia and, most of all, in China. It should be noted that there is structural imbalance in container transport between China and Western Europe and the USA. Traditionally, more loaded containers came from China to European and American ports than the other way around. This disproportion was usually neutralized by repositioning empty containers to Chinese ports. However, during the pandemic, this process has been disrupted. Container shippers have had to compete for containers, and as a result, freight rates have been considerably increased in container transport with China.

As a result of that turbulence, a huge increase in freight rates, fees for chartering container vessels and option costs for construction of new vessels (especially considering the necessity of adjusting propulsion systems to the strict requirements for lower emission) was recorded.

At the beginning of September 2021, Shanghai Containerized Freight Index, which presents freight rates in trade with Chinese ports, recorded an increase by

449% in comparison to the analogical period before the pandemic (September 2019) (Hellenic Shipping News 2021). The situation was directly translated into extraordinary revenues recorded by shipping companies. For instance, during the third quarter of 2021, A.P. Moller-Maersk recorded a gigantic increase in the net income for its operational activities: it reached the level of USD 5.461 billion in comparison to USD 0.947 billion recorded for the analogical period in 2020. The factors contributing to that situation were the following: freight rates going up to 3.651 USD/FEU (Q3 2021) on average from the level of 1.909 USD/FEU (Q3 2020), factors that lowered the profit, such as costs of heavy marine fuel, which increased in an analogical way—up to USD 1.4 billion (Q3 2021) from the level of USD 0.759 billion (Q3 2020). This reflected an increase in the prices of marine fuel by 74% on average (from USD 290 per metric tonne to USD 504 per metric tonne), considering the annual fuel consumption recorded by that shipping company at the level of 10.37 million metric tonnes in 2020.

The Covid-19 crisis has also affected shipyard industry. During the first quarter of 2020, shipyards obtained only 13 orders for new container vessels and in April 2020 they did not receive any orders at all (ISL Shipping Statistics and Market Review 2020, 5–6).

Vessel crews were also put in a very difficult situation because of the pandemic. In accordance with international regulations on employing crew members on seagoing commercial vessels, each month, about 150,000 seamen were supposed to go home and be replaced by other seamen. During the pandemic, it was impossible because of sanitary restrictions concerning the disembarkation of crew members and cancellation of air flights, which were usually taken by seamen to get home or to get aboard their vessels. Another obstacle to transferring crew members, who were frequently citizens of various countries, was the fact that consulates, where visas for seamen are usually issued, were closed. As a result, thousands of seamen were stuck on their vessels. Working at sea during the subsequent months without any possibility to see their families resulted in growing frustration, and it negatively affected the quality of seamen's performance (UNCTAD 2019, 49).

At the beginning of 2022, some other significant circumstances occurred, posing a threat of destabilization to European (and global) trade (including maritime containerized transport). On 24th February 2022, another war broke out when Russia invaded Ukraine (after the annexation of Crimea in 2014). Intensified war operations were launched, bringing destruction and severely affecting Ukrainian people, who bravely fought back against the invaders. The unprecedented Russian aggression (at present amounting to what is considered war crimes and genocide) faced the response of all the community of the democratic world in the form of various economic sanctions imposed on Russia. The prices of strategic raw materials rocketed skyward, the levels of strategic reserves in numerous European countries turned out to be too low and logistic supply chains to and from Russia were practically and largely stopped. The intention of the countries imposing sanctions on Russia (Europe and the United States of America) was to isolate that country politically and also to weaken it economically, causing its economic bankruptcy and, at the same time, providing strong political and economic support to Ukraine (including armament supplies). At the end of the third quarter of 2022, the military conflict remained unsolved. Assuming that the sanctions and the resistance of the Ukrainian army and civilians will bring positive results, and despite a partial (or in a less optimistic scenario, complete) annexation of Ukraine, Russia will be forced to withdraw from Ukraine, facing its own political and economic (and possibly military as well) bankruptcy.[9] Undoubtedly, in any predictable future, it should pave the way for the process of dynamic reconstruction and recovery of Ukraine (in a formula of a democratic country following the rule of law). The United States of America and European countries have already declared their financial, material, technological (knowhow) and political support on an unprecedented scale in history that can be counted in EUR billions. Most probably, a major part of the material support will be transported by sea to container terminals in Ukraine (Odessa,[10] Yuzhnyy, Chornomorsk—status as of 15th September 2022). It should bring international maritime containerized transport back to the Black Sea (on the day when the Russian invasion was started, the terminal Odessa was closed because the control over shipping on the Black Sea was seized by Russia, also because the Ukrainian port was blocked and the military threat at sea was very high).[11] The scale of the economic sanctions imposed on Russia and their influence on trade (including maritime transport of containerized cargo) between Russia (after the war) and EU countries and the USA are at present impossible to be assessed during the current stage of the conflict.

Considering the previously mentioned arguments, issues related to the development of the geopolitical situation and global trade conditions, it is possible to draw five general conclusions that determine future (post-Covid-19 and after the war in Ukraine) governance in the field of international (global) supply chains, directly affecting the shape of international maritime transport of containerized cargo:

- It is highly probable that a reconfiguration of supply chains and international exchange will take place, considering tension between China and the USA, the expansive policy of China in the development of the New Silk Road and searching of new locations for production and distribution based on reorientation of trade policy pursued by the "triad" entities. This will result in shortening and making current supply chains more flexible (new "geography" of connections).
- It is highly probable that production locations will be closer to potential sales markets. This will result in a reconfiguration of current production locations that have been so far dominated by the strong economy of China.
- It is highly probable that new guidelines concerning the amounts of reserves and flexibility of production and supply chains will be issued by the "triad" entities and countries directly affected by the war in Ukraine. This will result in an increase in reserves at the particular stages of logistics processes, in a change in the attitude toward obligatory (strategic) reserves and also in a necessity of increasing national reserves (which, in turn, will increase pressure on storage area, warehouse surface, cargo handling capacities of port terminals, etc.).
- Most probably, international logistic supply chains of transporting containerized cargo to Ukraine by sea will be ceased until the end of the military

operations; after that, a high dynamic in the turnover growth will be observed due to the implementation of reconstruction and recovery schemes.

- Most probably, international logistic supply chains of transporting container-ized cargo to and from Russia by sea will be constantly limited to their mini-mal levels, in accordance with the policy of sanctions becoming stricter and stricter. At present, the situation after the war is impossible to be predicted.

As presented previously, the situation in international maritime container transport has imposed some structural changes observed in relations among container opera-tors, particularly in their strategic alliances.

4.4 THE WORLD'S LARGEST CONTAINER OPERATORS IN MARITIME TRANSPORT

In global container shipping, a process of capital concentration (looking for some larger "critical mass") has been taking place. It is manifested by numerous mergers and take-overs and also by an increase in the leading container operators' market share. At the beginning of 2020, the world's three largest container operators in terms of the capacity (given in TEU) of their container vessels had 47.9% of the market share and the 10 largest operators—86%. In 2010, these shares were respectively 37.8% and 65.5% (Table 4.8).

TABLE 4.8
The World's Largest Container Operators (February 2020)

Container operator	Headquarters country	Vessels in total > 1,000 TEU		Share in % TEU	Average size of vessels in TEU
		Number	in 1,000 TEU		
MAERSK	Denmark	696	4,125.3	18.5	5,927
MSC	Switzerland	519	3,673.5	16.5	7,078
COSCO (OOCL)	China/Hong Kong	403	2,868.4	12.9	7,118
CMA CGM	France	426	2,596.3	11.6	6,095
Hapag-Lloyd	Germany	230	1,644.5	7.4	7,150
ONE	Singapore	218	1,553.2	7.0	7,125
Evergreen	Taiwan	195	1,270.9	5.7	6,518
Yang Ming	Taiwan	99	657.4	2.9	6,641
PIL	Singapore	101	411.0	1.8	4,069
HMM	South Korea	61	379.3	1.7	6,218
ZIM	Israel	66	314.7	1.4	4,769
Wan Hai	Taiwan	94	249.8	1.1	2,657
KMTC	Japan	57	151.8	0.7	2,663
SITC	Hong Kong	64	99.9	0.4	1,561
X-Press Feeders	Singapore	45	88.0	0.4	1,956
Other operators	–	1,033	2,233.9	10.0	2,163
Total	–	**4,307**	**2,2317.9**	**100.0**	**5,182**

Source: Data from ISL Shipping Statistics and Market Review (2020, 16).

Among the 15 largest container operators, there are four operators from Europe, three operators from Singapore and three operators from Taiwan. Chartered container vessels play a significant role for all the operators from the top 15 list. Three operators have container vessels of the largest average size at their disposal: the German Hapag-Lloyd (7,150 TEU), the Singaporean ONE (7,125 TEU) and the Chinese COSCO/OOCL (7,118 TEU). Nine out of the world's ten largest container operators belong to one of the three strategic alliances in container shipping: 2M Alliance, Ocean Alliance and The Alliance. These alliances have already dominated the global market of container shipping. They control 84% of the global container shipping capacity, and their share in the market of cargo shipping between Asia and Europe is 95% (ISL Shipping Statistics and Market Review 2020, 8).

Since April 2020, the dynamic situation resulting from changes in the volume of maritime traffic during the lockdown, from changes foreseen in the configuration of supply chains, from fluctuation in freight rates and from growing competitive pressure has established a new configuration of the current alliances. The HMM operator changed its previous interest in "the strategic cooperation" with 2M and has joined The Alliance. Instead of three Japanese maritime shipping operators, Kawasaki Kisen Kaisha, Ltd ("K" Line), Mitsui O.S.K. Lines, Ltd (MOL) and Nippon Yusen Kabushiki Kaisha (NYK), the Ocean Network Express (ONE)[12] has been established. The actual structure of the alliances is presented in Figure 4.1.

Alliances do not tie shipowners as closely as shipping consortia, but they concern providing numerous shipping services, and therefore, they give these agreements a global character. They allow partners to run joint operation of vessels, which involves the coordination of particular sailings, choice of ports and frequency of sailings. On the vessels belonging to the shipowners who are the members of an alliance, mutual freightage of available standard container slots takes place. The process is referred to as the slot exchange (Załoga 2014, 409–417, 58–59).

The process consists in sharing loading space: several shipowners divide the loading space of their vessels into slots, which are treated as the joint shipping capacity that is subsequently assigned to the particular partners who fill the slots with containers. Hence, shippers deal with various operators of slots on the same vessel, who can even compete with each other, offering different prices and quality service.

Considering the analysis presented previously, an operational model of maritime container transport in the world comes as another important question. Operators do

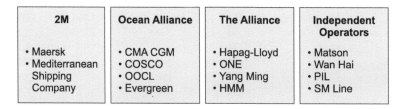

2M	Ocean Alliance	The Alliance	Independent Operators
• Maersk • Mediterranean Shipping Company	• CMA CGM • COSCO • OOCL • Evergreen	• Hapag-Lloyd • ONE • Yang Ming • HMM	• Matson • Wan Hai • PIL • SM Line

FIGURE 4.1 The Alliances of Shipping Operators (as of the Beginning of 2021).

Source: Adapted from www.geminishippers.com/hmm-enter-alliance/ (accessed: 27th April 2021).

not connect various ports directly; they operate on the basis of the hub-and-spokes system.[13] According to that system, it is possible to distinguish two basic types of shipping: ocean (line) shipping and feeder shipping. Ocean shipping includes (Urbanyi-Popiołek, ed. 2012, 69–70):

- regional (latitudinal) connections understood as shipping services offered between the most developed regions of the world—e.g., Asia–Europe–Asia;
- pendulum/shuttle connections understood as routes between highly developed regions connecting two container shipping routes, e.g., Asia–Europe–the eastern coast of North America–Europe–Asia; and
- round-the-world connections understood as shipping services connecting highly industrialized regions (centers); vessels operating on these shipping routes navigate in one direction, crossing three oceans—e.g., to the east–the eastern coast of North America–the Panama Canal–the eastern coast of North America.[14]

The actual geography of connections in maritime containerized transport is presented in Figure 4.2.

Table 4.9 presents the main container shipping services in the global maritime transport.

Ocean shipping is usually handled by large container vessels (ULCV, Panamax, often referred to as *mother vessels*), which—considering their parameters, mainly draft, length and beam—can call at very few ports located on the particular continents

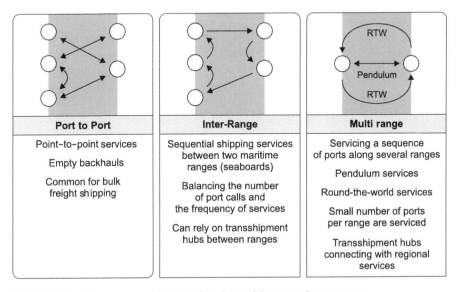

Port to Port	Inter-Range	Multi range
Point-to-point services	Sequential shipping services between two maritime ranges (seaboards)	Servicing a sequence of ports along several ranges
Empty backhauls		Pendulum services
Common for bulk freight shipping	Balancing the number of port calls and the frequency of services	Round-the-world services
		Small number of ports per range are serviced
	Can rely on transshipment hubs between ranges	Transshipment hubs connecting with regional services

FIGURE 4.2 The structure of connections in maritime container transport.

Source: Adapted from https://transportgeography.org/contents/chapter5/maritime-transportation/maritime-routes-types/ (accessed: 16th April 2021).

TABLE 4.9

The Number of the Main Container Shipping Services in the global Maritime Transport (as of December 2012)

Shipping routes	Number of shipping services
Far East–North America	73
North Europe–Far East	28
Far East–Mediterranean	31
North Europe–North America	23
Mediterranean–North America	21
Europe–Middle East/South Asia	40
North America–Middle East/South Asia	10
Far East–Middle East/South Asia	72
Australasia	34
East Coast South America	26
West Coast South America	48
South Africa	24
West Africa	60
Total	490

Source: Data from www.worldshipping.org, shipping services may be counted multiple times (for a higher number of routes), in accordance with the instructions provided by *World Shipping Council*, www.worldshipping.org/about-the-industry/global-trade/trade-routes (accessed: 17th May 2019).

(so called hub ports). Smaller ports located in the catchment areas of the particular hub ports are usually connected with them by a network of feeder services. The system based on the hub-and-spokes solution underlies transshipment processes.[15] The structure of this type of maritime connections in the global scale is presented in Figure 4.3.

Traditionally, shipowners of container vessels and containers used to focus exclusively on shipping containerized cargo by sea. Logistics, forwarding and cargo handling services were provided by companies from the outside of the maritime navigation field. At present, apart from shipping containers, large shipowner companies also offer cargo handling and forwarding services using their own organizational networks—for example, owning the Maersk Line, the A.P. Moeller Holding also owns APM Terminals, a company that deals with cargo handling, and the Maersk Logistics (Damco), a forwarding company. In the NYK Group, there is NYK, a shipowner, Terminals&Harbors, a company offering cargo handling services, and NYK Logistics, a forwarding company. Another solution is commissioning additional services to logistics departments or divisions that are incorporated into the structure of a particular company and treating them as subcontractors. Such a solution is applied by the following shipowners: Mitsui OSK Lines (MOL), China Shipping Container Lines (CSCL), Orient Overseas International (OOCL) and Hapag-Lloyd.

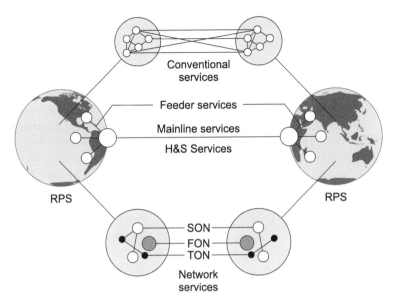

FIGURE 4.3   The network of global maritime connections (by type and connection structure).

Source: Adapted from Rodrigue (2017, 37, 41).

Legend:

RPS—Regional Port System
H&S—Hub and Spokes Services
FON—First order network
SON—Second order network
TON—Third order network

The main reason for shipowners to extend their shipping services by land transportation is to maintain control over the process of shipping containers along their entire shipping routes and to improve the way how their container stock is used. An insufficient number of empty containers has resulted in some problems in the access to maritime transport services, and it has also become one of the reasons for an increase in freight rates. Hence, shipowners have decided to get involved into the organization of land shipping and cargo handling services in order to optimize their container turnover and to reduce costs related to those processes.

Expanding the range of operations undertaken by maritime operators of container transport by additional fields of business activities, functioning under the structure of international alliances and pursuing individual developmental strategies allow the authors to point out several leading entities, such as the following: A.P. Møller—Mærsk, a Danish holding; Mediterranean Shipping Company (MSC), a Swiss-Italian group; COSCO Shipping Lines Co, a Chinese company; CMA CGM, a French shipowner; Hapag-Lloyd, a German company; Ocean Network Express—ONE, a Japanese group; and Evergreen Marine Corporation (EMC), a container shipowner from Taiwan, all of which take the leading positions in the world's shipowner top list.

4.5   THE CHARACTERISTICS OF THE SELECTED CONTAINER SHIPOWNERS

A.P. MØLLER—MÆRSK

Since the mid-1990s, A.P. Møller—Mærsk, a Danish group (commonly known as Maersk), has been the largest container vessel fleet operator in the world. At the beginning of 2020, the company had 696 container vessels of the capacity of 4.1 million TEU, which accounted for 18.5% of the world's container tonnage. The fleet includes one of the largest container vessels in the world, the *Madrid Maersk*, which is 400m long, 59m wide, and its capacity is 20,568 TEU. In the *Global 500* ranking published by *Fortune*, an economic magazine, the ᴹaersk group takes the 305th position. The Maersk company was established in 1904 by Captain Peter Møller—Mærsk and his son, Arnold Peter Møller as Dampskibsselskabet Svendborg. Today, it is a global conglomerate with its headquarters based in Copenhagen, which employs 80,000 workers in 130 countries. Apart from container shipping, the company deals with maritime transport of bulk cargo, operation of container terminals and activities in the energy sector. Within the group, the entity responsible for container shipping and operation of bulk carrier vessels is Maersk Line, a subsidiary company, along with Hamburger Süd, a company whose vessels call at over 300 ports in over 100 countries. Before 2010, Maersk commissioned construction of vessels mainly to a shipyard in Odensee and later on to some Asian shipyards as well. In October 2020, the company placed orders for 14 new container vessels. The majority of vessels operated by Maersk (382 vessels) are chartered ships (Statista 2021e).

The strategic container hubs for the vessels of the Maersk group are the following ports: Rotterdam Maasvlakte II, Algeciras, Tangier, Tangier-Med II, Port Said, Salalah, Tanjung Pelepas and Bremerhaven. The capacity of containers loaded on the vessels operated by Maersk in 2019 was 13.3 million TEU (Table 4.10). The highest numbers of containers are transported by the vessels of Maersk along the ocean routes north–south (47.8%).

TABLE 4.10

The Volume of Ocean Container Shipping Achieved by the Vessels of the Maersk Group in the Years 2018–2019 (in TEU Thousand)

Transport direction	2018	2019	Change in %
East–West	4,186	4,100	−2.1
North–South	6,450	6,362	−1.4
Interregional	2,670	2,834	−6.1
Total	13,306	13,296	−0.1

Source: Data from Maersk (2020, 40).

MEDITERRANEAN SHIPPING COMPANY (MSC)

In accordance with the data from the beginning of 2020, the second position on the list of the global container shipowners is taken by a Swiss-Italian Mediterranean Shipping Company (MSC), with its headquarters based in Geneva. It was established in 1970 by an Italian Captain, Gianluigi Aponte. At present, the MSC employs 70,000 workers (including passenger vessels), and it has 519 container vessels (larger than of the capacity of 1,000 TEU) of the total capacity of 3,673,000 TEU, which accounts for 16.5% of the global fleet of container vessels. The fleet of the company includes two large container vessels, namely the MSC *MSC Gülsüni* and the *MSC Samar* of the cap¾ty of 23,756 TEU. About 3/4 of the fleet operated by the MSC are chartered vessels. Cargo shippers can contact the company through almost 500 MSC offices in 155 countries. The MSC container vessels provide services along over 200 container shipping routes, they call at 315 ports (including Gdynia) located on all the continents and they transport containers of the capacity of 21 million TEU (MSC Mediterranean Shipping Company 2021). The MSC is also a recognized operator of passenger vessels and container terminals. It also operates as a large logistics company.

COSCO SHIPPING LINES CO

Considering the size of the fleet given in TEU, the third position on the list of the largest container shipowners is taken by COSCO Shipping Lines Co., with its headquarters based in Shanghai, which is a part of the Chinese state-owned COSCO Shipping Group. The shipowner provides container shipping services to Chinese ports, international container shipping and related services in the field of maritime trade. At the end of 2020, the company had 423 container vessels of the capacity of 2.3 million TEU (the entire COSCO Shipping Group operated 536 container vessels of the capacity of 3.1 million TEU). The COSCO Shipping Lines Co. transported containers along 265 international shipping routes, calling at 354 ports in 105 countries and providing shipping services along 134 Chinese routes (Cosco Shipping Lines Co. 2021). In 2019, the volume of cargo shipping reached the level of 18.8 million TEU, and the revenues obtained from the services provided along all the routes reached the level of 982 million RMB (Table 4.11).

CMA CGM

The fourth biggest fleet of container vessels belongs to a French shipowner, CMA CGM, with its headquarters based in Marseille. The CGT company waˢ established in the mid-19th century, and in 1998 it was privatized and taken over by the CMA shipowner. Since then, the company has been expanding and it has become the largest shipping enterprise in France. Now it operates under the name of CMA CGT. The CMA CGT group has been vastly expanded, having taken over several other companies: Australian National Lines (1998); Delmas, a French shipowner (2005); LCL Logistix, an Indian logistics company (2015); and NOL, a Singaporean container shipowner (2019).

TABLE 4.11

Shipping and Revenues of the COSCO Shipping Lines Co. Broken by the Main Cargo Shipping Routes

Shipping routes	2018 mln TEU	2018 mln RMB	2019 mln TEU	2019 mln RMB
Trans-Atlantic	2.865	236	2.670	234
Asia–Europe	3.173	184	3.484	205
Asia + Australia	4.746	185	4.899	210
Other international	1.832	124	1.996	149
China	5.749	118	5.736	123
Total	18.366	847	18.785	923

Source: Data from Cosco Shipping Holdings Co. Ltd. (2020, 23).

In 2019 the CMA CGT group operated 489 vessels to transport containers of the capacity of 2,705,000 TEU; 1/3 of these vessels were the company's own ships and 2/3 were chartered vessels. Today, the company is present at 420 ports based in 160 countries and is available to its customers through a network of 755 agents and 750 warehouses. The company handles 200 shipping routes, shipping containers of the capacity of 21.6 million TEU annually (CMA-CGM 2021).

HAPAG-LLOYD

Based in Hamburg, Hapag-Lloyd is the largest shipping organization in Germany. It specializes in shipping containerized cargo by sea. The group also runs operations in the field of logistics and house-to-house cargo deliveries. The largest operating company in the group is Hapag-Lloyd AG, which provides 99% of the turnover of the whole group. It has the following subsidiaries: Hapag-Lloyd Rotterdam, Hapag-Lloyd-Antwerpen, Hapag-Lloyd Denmark (Holte) and Hapag-Loyd Poland (Gdańsk). Hapag-Lloyd was established in 1970 as a result of a merger of two German ship-owners of well-established tradition: Hamburg—Amerkanische Packetfahrt AG (HAPAG) in Hamburg and Nordeutscher Lloyd in Bremen. The first of the mentioned companies was established in the market of maritime transport in 1847, whereas the latter one in 1857. At the beginning of 2020, Hapag-Lloyd employed 13,200 workers, including 11,200 employees on land and 2,100 employees at sea. The company had its representation in 388 cities located in 129 countries, and its main offices in Hamburg, Piscataway (the USA), Genoa, Dubai, Singapore and Valparaiso. The company had 234 container vessels of the capacity of 1.7 million TEU, including 50 vessels of the capacity over 50,000 TEU and a huge stock of containers of the total capacity of 2.7 million TEU. Hapag-Lloyd handled 121 container shipping routes, providing connections to 600 ports based on all the continents. In 2019 the company's vessels transported containers of the capacity of 12 million TEU (Hapag-Lloyd 2021). The largest share in cargo transported by the company is taken by food products (16%), plastic goods (14%) and chemical products (14%), as seen in Table 4.12.

TABLE 4.12

The Structure of Containerized Cargo Transported by Hapag-Lloyd in 2018

No.	Goods	Share in %
1	Food products	16
2	Plastic goods	14
3	Chemical products	14
4	Paper and wooden goods	9
5	Machinery and technical appliances	9
6	Raw materials	8
7	Textiles	8
8	Automotive parts	6
9	Electronic goods	5
10	Furniture	5
11	Others	7
	Total	100

Source: Data from Hapag-Lloyd (2020a, 72).

TABLE 4.13

Container Shipping and Revenues Obtained by Hapag-Lloyd for Container Shipping in 2019

Shipping routes	Number of routes	Shipping in TEU thousand	Revenues in EUR million
Trans-Atlantic	21	1,960	2,432
Trans-Pacific	17	1,945	2,291
Europe–Far East	8	2,327	1,892
Middle East	9	1,391	925
Intercontinental Asia	24	900	435
Latin America	25	2,837	2,922
Mediterranean Europe–Africa	17	676	632
Total	121	12,037	12,608*

Note: * The difference results from the fact that other revenues have not been taken into consideration.

Source: Data from Hapag-Lloyd (2020a, 84–85).

Hapag-Lloyd transports containers along numerous shipping routes, the most important of which are presented in Table 4.13.

The largest and highest revenues are achieved by Hapag-Lloyd in container turnover in Latin America and on the trans-Atlantic route.

Ocean Network Express—ONE

Considering the size of its fleet of container vessels, ONE is the sixth ship-owner in the world, with the share of 6.5% in the global tonnage. The company was established in 2017 as a result of a merger of three Japanese shipowners: K LINE, MOL and NYK. ONE started its business operation in 2018. It belongs to a Japanese holding, with its main headquarters based in Singapore. Additionally, there are five regional offices: in Hong Kong, Singapore, London, Richmond (the USA) and Sao Paulo. ONE provides employment to 8,000 workers and has 224 vessels of the shipping capacity of 1.59 million TEU. The company handles over 130 container shipping routes, providing connections to over 130 seaports based on all the continents. ONE offers shipping services, including transportation of all types of general cargo, cooled products, hazardous and oversized goods (ONE Ocean Network Express 2021). Along with a Taiwanese shipowner, Yang Ming, and a Korean container shipowner, Hyundai Merchant Marine (HMM), ONE forms The Alliance consortium.

During the first financial year of its business operation, from April 2018 to March 2019, ONE reported loss, but during the three subsequent quarters of 2019, the company achieved profit at the level of USD 131 million at the turnover at the level of USD 8.9 billion (DVZ Deutsche Verkehrs Zeitung 2021).

Evergreen Marine Corporation (EMC)

A Taiwanese container shipowner, Evergreen Marine Corporation (EMC), based in Taipei, is a part of a large maritime corporation, Evergreen Group, which also operates a number of container terminals in the world. At the beginning of 2020, the fleet of container vessels belonging to EMC shipowner companies—namely to Evergreen Marine Corp. (Taiwan), Italia Marittima, and Evergreen Marine (Singapore)—included 197 vessels of the loading capacity of 1,271,000 TEU, 88 of which were chartered vessels (ISL Shipping Statistics and Market Review 2020, 16) that accounted for the capacity of 656,000 TEU (51%). The EMC shipowners operate along 240 container shipping routes, providing connections to 240 ports. They offer services along the following maritime shipping routes:

- Far East–North America
- Far East–North Europe and Mediterranean countries
- Europe–East Coast of the USA
- Far East–Latin America
- Far East–Africa
- Asia–Middle East, Indian subcontinent, Australia
- Intra-Europe and Mediterranean countries (Evergreen 2021)

Considering competitiveness of shipowners, elements such as assets, capital or economic accounts become crucial. In order to analyze these parameters, Hapag-Lloyd has been selected here as an example.

4.6 THE ASSETS, CAPITAL, REVENUES AND COSTS OF THE SELECTED CONTAINER SHIPOWNER—HAPAG-LLOYD

An annual financial statement for the particular year presented by a company usually comes as a reliable source of data concerning that company's assets, liabilities, revenues and costs. The analysis of the basic economic aspects of the Hapag-Lloyd group is based on its financial statement for 2019, as seen in Table 4.14 (Hapag-Lloyd 2020b, 154–155).

The most important elements of Hapag-Lloyd's property are fixed tangible assets, which account for 62.1% of the group's assets. The share of intangible and legal assets is also relatively high because it accounts for 20.5% of the balance sheet total. The share of the financial components of the property is low as they account for 2.3% of the company's assets. Hapag-Lloyd funds its operation with its own financial means in 40.9%, where the initial capital accounts for only 1.1% of all the liabilities. The share of the foreign capital in the liabilities total is 59.1%, where long-term liabilities account for 34.5% of all the fund sources and short-term liabilities account for 24.7%.

In accordance with the profit and loss account published by Hapag-Lloyd, its revenues from the business operation in 2019 were EUR 12 608 million and were higher by EUR 999 million than in 2018. They depend on the capacity of the containers that have been transported and on the freight rates for shipping containers along the particular maritime routes. The highest revenues were achieved along the Pacific route (1,389 USD/TEU), the Atlantic route (1,318 USD/TEU) and the route covering ports of Latin America (1,152 USD/TEU). The revenues of Hapag-Lloyd that were achieved from its financial operations were minimal. The highest position among costs was taken by the costs related to container shipping, which reached the level

TABLE 4.14
The Balance Sheet of the Hapag-Lloyd Group for 2019

Assets	EUR million	Liabilities	EUR million
Fixed assets,	13,811.8	Equity capital	6,620.6
including:	3,317.6	including:	175.8
intangible and legal assets	10,064.9	initial capital	2,637.4
fixed tangible assets	429.3	reserve capital	3,793.4
long-term receivables		other equity capital	
Current assets	2,388.6	Foreign capital	9,579.8
including:	511.6	including:	5,586.2
cash	248.5	long-term liabilities	3,993.6
reserves	1,628.5	short-term liabilities	
short-term receivables			
Total assets	16,200.4		16,200.4

Source: Data from Hapag-Lloyd (2020b, 154–155).

of EUR 9,707 million (16.8% of all the expenditures for shipping operations). The structure of those expenditures was the following (Hapag-Lloyd 2020b, 86):

- ship fuel—EUR 1,626 million (16.8%)
- handling and haulage (charges for shipping containers on land and handling containers at a container terminal)—EUR 4,923 million (50.7%)
- costs related to vessel operation (port fees and charges for using canals, costs of ship repairs, costs related to slot chartering operations)[16]—EUR 1,968 million (20.3%)
- costs related to containers (leasing, repairs, cleaning) and their repositioning—EUR 1,205 million (12.4%)
- personnel costs—EUR 683 million (7.1%)
- depreciation costs—EUR 1,174 million (12.1%)

The development of the fleet of container vessels, the structure and the range of operations undertaken by the main stakeholders in the field of maritime container shipping—shipowners (and their alliances)—directly determine other parameters of the discussed market, namely the size and the geographic structure of international container shipping by sea.

NOTES

1. Commonly referred to as mega-container ships, which are adjusted to transport ISO containers only.
2. The size of Panamax vessels allows them to sail through the Panama Canal.
3. Feeder container vessels are used for short-sea shipping.
4. See more in Przepisy klasyfikacji i budowy statków morskich. Część I Zasady klasyfikacji (2021, 9–16); Neider (2012, 37–39; Rydzkowski and Wojewódzka-Król, eds. 2006, 123–124).
5. See more in Miler (2016a, 48–50).
6. Typical sto-ro vessels are equipped neither with external nor with internal ro-ro ramps. They are equipped with side doors with lifts/side loaders; a port machine that is operated along the wharf puts a piece of cargo (a bale of paper of a pallet, etc.) on a vessel outer platform extended from the hull, with a conveyor belt. The platform of the cargo lift is stopped at the level of the vessel outer platform (as its extension), and the cargo is moved to its destination deck level, where it is picked from the platform by a forklift or a stacker operated in the hold/particular deck. Then the cargo is moved to its destination place.
7. At the beginning of 2020, vessels launched before 2000 accounted for only 5.5% of the container vessel tonnage (given in TEU) (ISL Shipping Statistics and Market Review 2020, 29).
8. The authors' own calculations based on the data provided by UNCTADstat.
9. Obviously, considering the current knowledge about the conflict and assessment of risk, some other less optimistic scenarios might occur in the development of the military situation in the discussed region, such as further internationalization of the conflict and its escalation in terms of its military aspect, humanitarian crisis and military involvement of other countries.
10. OCT—Odessa Container Terminal (a subsidiary of a German holding group Hamburger Hafen und Logistik AG (HHLA). It is the largest container terminal in Ukraine. In 2021,

390,000 TEU were handled there, and in the entire port of Odessa, container handling operations reached the level of 671,000 TEU.

11. On 24th February 2022, the owner of the OCT container terminal in Odessa, the HHLA operator of the port of Hamburg, decided to close the terminal and to send its employees home. The 44 Project company informed that there were two vessels stuck at the closed terminal. According to the data provided by the 44 Project, starting from the beginning of the crisis between Russia and Ukraine in December 2021, the terminal in Odessa has been suffering from its impact. In February 2022 the volume of cargo that was transported to the port was by 30% lower than in December 2021. Three other large logistics companies—namely DB Schenker, Hapag-Lloyd and Maersk—made similar decisions and closed their offices in Ukraine and stopped their operation in that country.

12. According to the founders, in the new structure the merger will allow ONE to choose and to apply the best practices taken from each of three partners and to use the joint shipping capacity of their container vessel fleets at the level of 1,440,000 TEU and also to optimize over 85 shipping routes handled by the new carrier. The vessels belonging to ONE have got new colors: their hulls and containers are painted white and pink. It is supposed to bring their greater recognisability in the market.

13. This solution has been adapted from air shipping.

14. Widespread implementation of post-Panamax container vessels that, considering their size, cannot use the Panama Canal has resulted in the further development of this type of navigation and also in a simultaneous increase in the competitiveness for round-the-world services.

15. At present, transshipment accounts for 26% of the container turnover at sea ports in the world—data from Balticon (2010, 7).

16. The entity chartering slots operates as an independent shipping line, and it uses its own equipment and bills of lading. It has also got its own account at the port, and the port issues invoices directly to that entity for commercial fees.

5 The Size and Geographical Structure of International Container Shipping in Maritime Transport

5.1 THE GLOBAL STOCK OF CONTAINERS APPLIED IN MARITIME TRANSPORT

According to Drewry,[1] a British maritime research center, in 2019 the global stock of containers applied in maritime transport represented the capacity of 42 billion TEU, 7% of which accounted for refrigerated containers (Triton 2020, 5–6). The structure of the container stock by the size of containers is as follows (Forschungs Informations System 2021):

- 40′ high-cube containers—40%
- 20′ general-purpose containers—30%
- 40′ general-purpose containers—13%
- regional containers—4%
- 45′ containers and others—1%

The number of sailings covered by containers is ten times higher than their number. The average time for operating a container in maritime transport is 15 years. After that, containers can be still used for storage of goods. Production of containers is dominated by China, where 85% of the global container supply comes from. The largest cargo container manufacturer is the China International Marine Containers (Group) Ltd (CIMC),[2] with its annual production capabilities of 2 million TEU.

The cost of a new cargo container depends, first of all, on its type and size. Special-purpose containers, such as reefers, are more expensive than general-purpose containers. Secondhand containers are usually cheaper and there is a lot of them in the market. The price of a new 20′ container is approximately USD 3000; a second-hand 20′ container costs about USD 2000; a new 40′ container is about USD 4500 and a second-hand 40′ container is USD 2200. The value of all cargo containers manufactured in the world in 2019 was USD 8.7 billion (Container xChange 2021). The largest container manufacturers are presented in the following table (Table 5.1).

TABLE 5.1

The Largest Cargo Container Producers in the World (2019)

Manufacturer	Headquarters	Annual production capabilities in thousand TEU
CIMC (China International Marine Container Group Co.)	Shenzhen, China	2,000
Singamas	Shanghai, China	480
CXIC (CXIC Group Containers Co.)	Changzhou, China	800
COSCO Shipping	Shanghai, China	500
CEC (China Eastern Containers)	Shanghai, China	150
W&K Container Inc.	California, USA	no data provided
Daikin Industries	Osaka, Japan	no data provided
Maersk Container Industry	Copenhagen, Denmark	no data provided
TLS Offshore Containers International	Singapore	no data provided
YMC Container Solutions	East Yorkshire, UK	no data provided
DCM Hyundai Limited (DHL)	Faridabad, India	no data provided

Source: Data from Container xChange (2021).

A vast majority of containers in the world is owned by container shipowners and leasing companies that lease their containers to shipowners. Only a small percentage of the global container stock belongs to shippers and forwarders. In 2019, container leasing companies owned 52.4% of the global stock of containers, which accounted for container capacity of 22.1 billion TEU (Triton 2020, 5–6). Shipowners lease containers because container shipping is characterized by high fluctuation, and therefore, there might be some difficulties in forecasting the demand for containers on a particular date and at a particular port. Container leasing comes to shipowners as an alternative funding source for container operation. It allows them to be flexible in adjusting the size of the fleet of container vessels they want to engage into operation and the structure of the container stock to the current demand for shipping containers by sea. Leasing contributes to better availability of containers at various ports where they are needed, without the necessity of keeping reserves. It also contributes to a decrease in expensive transportation of empty containers. All these factors contribute to the improvement in the efficiency of the fleet of container vessels.

The market of container leasing is dominated by five large leasing companies that own 86.3% of its share (Table 5.2).

There are three basic forms of container leasing in the world (Marciniak-Neider and Neider, eds. 2014, 310):

- Long-term leasing—it accounts for almost 70% of leasing transactions, the lessee takes containers for a longer period of time (48 months on average); the lessee is responsible for maintaining the containers in good condition and for repositioning empty containers.
- Master leasing—it involves leasing containers for a higher number of sailings.

TABLE 5.2

The Largest Companies That Lease Maritime Containers in the World

No.	Name of the leasing company	Headquarters	Container stock in TEU million	Share in the market of container leasing in %
1	TRITON International Limited	Bermuda	6.1	27.6
2	Bohai Group	Hong Kong	4.0	18.1
3	Floreas Container	Hong Kong	3.7	16.7
4	Textainer Group	San Francisco	3.6	16.2
5	Sea Cube Container	New Jersey	1.7	7.7

Source: Data from: www.tritoninternational.com; www.bohaiholding.com/en; www.florens.com/#/; www.textainer.com; https://seacubecontainers.com (accessed: 12th July 2021).

- Short-term leasing (spot leasing)—it involves leasing for a particular sailing; leasing rates undergo high fluctuation. The lessee is not responsible for delivering containers to the depot they have been taken from.

Leasing rates mostly depend on prices for new containers, which in turn depend strictly on the price of steel on the international market. Furthermore, leasing rates are also affected by the demand and supply on the market of container leasing. However, the way in which leasing rates respond to any changes in container prices and demand for containers is gradual rather than sudden. Usually, leasing rates are changed after the previous leasing agreement is terminated and at the moment when a new agreement is signed.

At present, the stock of containers operated in the world is characterized by high technical adjustment to cargo that is going to be transported considering its suitability for containerization.

5.2 SUITABILITY OF CARGO FOR CONTAINERIZATION IN INTERNATIONAL TRADE

If technical requirements stated for transporting containerized cargo by sea are met, namely, when there is a fleet of container vessels capable of transporting the cargo and port terminals are accessible, then the size and directions of international container shipping by sea depend on exporters and importers' demand for this kind of shipping. Some types of exported or imported cargo cannot be transported in containers. It depends on the physical and economic suitability of the cargo for containerization. There are three types of suitability for containerization (Miotke-Dzięgiel 1996, 5–8; Wiśnicki, ed. 2006, 31–35):

- natural suitability
- physical suitability
- economic suitability

Considering their natural (physical and biological) characteristics, some types of cargo, such as living animals or inflammable substances, cannot be transported in containers. Other types of goods that are characterized by low suitability for transportation, such as perishable and fragile goods or cargo that is vulnerable to weather conditions, can have their suitability for transportation improved by being transported in containers, which become additional packaging that can protect the cargo during the shipping and can provide, for example, optimal temperature. Considering some technical reasons, oversized cargo, the weight and size of which exceed standard dimensions, cannot be transported in containers because it is very large, heavy or spatial.

The economic suitability of cargo for containerization depends on its value, costs related to its indispensable packaging, transport costs in the entire transport chain and the size of the load weight. The economic suitability of cargo increases if the value of transported goods is higher, the total costs of transporting the goods in containers are lower and also if the load weight of the goods transported is heavier. Furthermore, if there are any alternative possibilities of transporting the goods, they should be taken into consideration before a decision about transporting them in containers is made.

General cargo is the most suitable for containerization. After the introduction of special-purpose containers into operation, the share of containerized general cargo in global shipping of general cargo has been systematically growing, and it has now reached the level of 90% approximately. Solid mineral fuels and crude oil are not suitable for container transportation. It might be surprising, however, that some dry bulk and semi-bulk cargo is transported in containers. Table 5.3 presents the containerization suitability factors for ten groups of cargo, in accordance with the classification stated for the requirements of the National Transportation Statistics.

In container shipping, it is very important to select a proper size of containers for a particular type of goods. In order to do it, shippers usually apply a container stowage factor that should be adequate to a cargo stowage factor. The container stowage factor is calculated as the quotient of its capacity given in m³ to the cargo mass, given in tonnes, that can be loaded into a container. Similarly, the cargo stowage factor is calculated as the quotient of its cubic capacity, given in m³, to the weight given in tonnes; for example, the stowage factor for shoe cardboard boxes is 4.25, and for reeled electric cables it is 0.99.

Containerization suitability in maritime transport comes as one of the factors that can affect foreign trade in the particular countries. It is the highest in the countries where maritime transport plays an important role in handling goods in foreign trade operations and where the share of highly processed goods is high in export—it is usually combined with importing components required for their production and importing labor-consuming consumer goods. Japan and South Korea are the best examples of the countries where such conditions can be observed. Considering foreign trade among the EU countries, where highly processed goods dominate, shipping containers by sea plays a less important role because distances between trade partners are usually shorter (it mainly involves short-sea shipping).

TABLE 5.3

The Containerization Suitability Factors for the Particular Cargo Groups

Group	Specification	Containerization suitability factor
0	Plant and forest products, living animals (cereals, vegetables, fruit, wood, textiles, animal and plant raw materials)	0.82
1	Food and animal fodder (food products, animal fodders, oil seeds, fat)	0.85
2	Solid mineral fuels (coal, coke, wood)	0.00
3	Oil and oil-derived products	0.12
4	Ores, metal waste and scrap metal	0.20
5	Metal products	0.20
6	Raw and processed minerals, construction materials (cement, lime, industrial construction materials)	0.30
7	Fertilizers	0.80
8	Chemical products (chemicals, chemical substances, paper mass, wastepaper)	0.80
9	Machines, transportation equipment, industrial goods (engines, machinery, cars, metal products, glass, pottery, leather, miscellaneous goods)	0.75

Source: Data from Wiśnicki, ed. (2006, 34).

However, it is only after the analysis of the economies of scale related to the volume and directions of international container shipping by sea when the fundamental significance of that process in global economy can be discussed.

5.3 THE VOLUME AND DIRECTIONS OF GLOBAL FLOWS IN MARITIME CONTAINER TRANSPORT

The volume of global container shipping by sea used to grow fast until 2008 when, as a result of the global financial and economic crisis, it suffered a considerable decline (Figure 5.1). During the subsequent years, the volume of global container shipping by sea started to grow again but at a slower rate; in 2019, it grew only by 1.1% in comparison to the previous year, reaching the level of USD 151.9 million. In 2017, the value of goods transported in containers by sea was USD 6.4 billion, which accounted for 36% of the global export value (Grzelakowski 2019). According to Clarkson,[3] in 2020, container vessels in the world transported goods worth over 60% of the value of the global maritime trade (Clarksons 2021). Among those goods, high-value products dominated: household electronic goods, automotive accessories, machinery, textiles, chemical substances, food and others. At the same time,

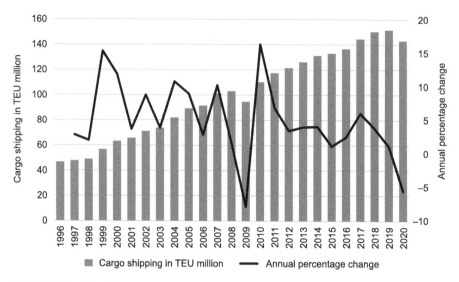

FIGURE 5.1   Container shipping in the world in the years 1996–2020 (stated in TEU and the annual percentage change).

Source: Data from UNCTAD calculations, based on data from MDS Transmodal (2020).

containers were more and more often used for transportation of bulk cargo, such as scrap metal, wastepaper or garbage.

Of all global container shipping, 39.2% of takes place from the west to the east and from the east to the west (Table 5.4). Containerized cargo transport is implemented along three important shipping routes (Clarksons 2021):

- the trans-Pacific route: East Asia–North America and North America–East Asia, along which containers of the total capacity of 26.8 million TEU were transported in 2019
- the route from East Asia to North to Mediterranean Europe and from North and Mediterranean Europe to East Asia, along which containers of the total capacity of 24.7 million TEU were transported
- the trans-Atlantic route, which connects Europe with North America, along which containers of the total capacity of 8 million TEU were transported

Intraregional shipping plays a significant role in the world's container trade. In 2019, intraregional shipping reached the level of 41.2 million TEU. Apart from the main shipping route between the East and the West, along which flows of cargo are sent between East Asia, Europe, North America and West Asia and the Indian subcontinent, containerized cargo shipping reached the level of 19.9 million TEU. Furthermore, along the north–south route (Europe, North America–Latin America, Oceania and Sub-Saharan Africa) containerized cargo shipping was recorded at the level of 12.0 million TEU and shipping along the south–south route (East Asia, Latin

TABLE 5.4

Maritime Container Shipping along the Three Most Important Routes in the Years 2014–2018 (Stated in Million TEU)

	Trans-Pacific route			Europe–Asia route Asia–Europe route			Trans-Atlantic route		
Years	East Asia–North America	North America–East Asia	Total	Europe–East Asia	East Asia–Europe	Total	North America–Europe	Europe–North America	Total
2014	16.2	7.0	23.2	6.3	15.5	21.8	2.8	3.9	6.7
2015	17.4	6.9	24.3	6.4	15.0	21.4	2.7	4.1	6.8
2016	18.2	7.3	25.5	6.8	15.3	22.1	2.7	4.3	7.0
2017	19.4	7.3	26.7	7.1	16.4	23.5	3.0	4.6	7.6
2018	20.8	7.4	28.2	7.0	17.3	24.3	3.1	4.9	8.0
2019	20.0	7.0	26.8	7.2	17.5	24.7	2.9	4.9	7.5

Source: Data from UNCTAD (2020, 15).

TABLE 5.5

The Global Containerized Cargo Shipping by Sea in 2019

	2016	2017	2018	2019
Main shipping routes		Stated in million TEU		
Main east–west routes	54.6	57.7	60.5	59.5
Other east–west routes	17.9	19.0	19.0	19.9
North–south	11.1	11.8	12.0	12.0
South–north	16.3	14.6	18.9	19.4
Intraregional shipping	36.6	38.8	40.0	41.2
The world	136.6	144.8	150.3	161.9

Source: Data from UNCTAD (2020, 14).

America, Oceania, Sub-Saharan Africa) reached the level of 19.4 million TEU as seen in Table 5.5 (UNCTAD 2019, 13–14).

In the years 2016–2019, containerized cargo shipping reached the highest volumes along the south–south route (19%) and along the main east–west route (10%). Asian countries are those that contribute the most to containerized cargo shipping by sea: China, South Korea, Japan and others. The share of Asia in that shipping is over 60%, whereas the share of Europe does not exceed 20%. Another big player is the USA. In Europe, containerized cargo shipping by sea competes with shipping by land (road and railway transport), considering relatively short distances to be covered. Containerized cargo shipping by sea among the EU countries is usually implemented by short-sea shipping, which covers European sea areas between various ports of the European Union and also ports of the third countries located at the European and adjacent seas.[4]

Similarly to the predictions referring to directions for the development of the container vessel fleet and to the predictions on directions for transportation at the beginning of 2021, it is now impossible not to refer to the reality of global pandemic conditions. The Covid-19 pandemic has already resulted in a drastic fall in container turnover in the world. Supply chains from Asia to West Europe and to the USA have been broken, and in China, a shortage of empty containers has been observed. Furthermore, shipowners have withdrawn the container vessel tonnage, which has exceeded the demand for container shipping services.

All these facts have resulted in a considerable increase in freight rates charged for shipping containerized cargo. As mentioned in the previous chapter, the increase in transport costs will accelerate the reorganization of supply chains in a longtime perspective, to make them shorter and more diversified. This, in turn, will decrease the demand for containerized cargo shipping services in the global scale. The industrial revolution 4.0 will affect maritime container transport in the same way. As soon as the production of labor-consuming goods is largely automated, moving it to some remote countries offering low labor costs will not be an option anymore, along with transporting such goods to Western Europe or to the USA.

China will still remain the most important market of container shipping in the future; however, the significance of other regions in the world will become stronger—for example, of Nigeria, the population of which will increase from the current level of 175 million up to 440 million in 2050, according to some forecasting provided by the United Nations. It is also predicted that the range of goods transported in containers will be extended. It will be possible a result of better adjustment of goods and packaging made by their manufacturers to the requirements of containerization. A graphic example is the Swedish paper industry, in which the size of paper bales has been already adjusted to the container dimensions. Similarly, it is possible to increase the volume of shipping lumber in containers. Surely, in the future, transportation of scrap metal, waste, waste paper, artificial fertilizers, peat and other numerous products in containers will be increased to some higher levels than today (Nowak 2019).

The IT revolution will become a significant factor to all changes observed in the field of maritime container transport. Customers will be acquired and handled more often with the use of internet platforms (e.g., e-commerce or blockchain technology), and the process of managing loaded and empty containers in the world will be improved (e.g., through process management in the cloud).

Another factor restraining possibilities of further development to maritime transport of containerized cargo refers to some limitations imposed on developmental capabilities of seaports (and container terminals).

5.4 THE VOLUME AND GEOGRAPHICAL STRUCTURE OF THE GLOBAL CONTAINER TURNOVER AT SEAPORTS

Shipping containers by sea starts and ends at a seaport, which is adjusted to loading and unloading containers. Hence, the global network of seaports (with modern container terminals) (Miler 2016a, 33) becomes especially important to optimization processes in maritime transport of containerized cargo. In mid-2020, there were 939 ports in the world, which means that there were 105 ports more than in 2006, where container vessels called regularly. These ports were connected by 12,748 direct shipping lines. Cargo shipping to other seaports had to be transshipped via a third port or several cargo handling ports. Figure 5.2 presents seaports with the highest number of regular shipping connections.

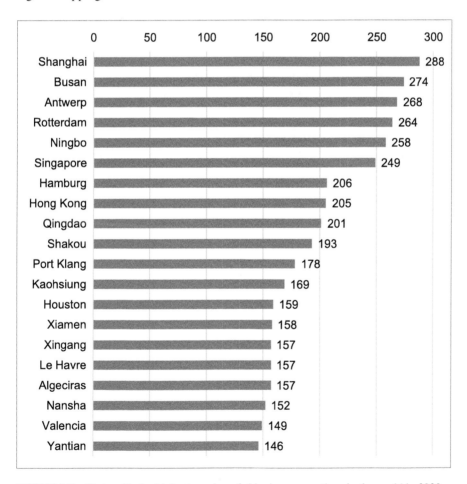

FIGURE 5.2 Ports with the highest number of shipping connections in the world in 2020.

Source: Data from Hoffmann and Hoffmann (2020).

The seaports characterized by the highest numbers of shipping connections in Asia are Shanghai (288), Busan (274), Ningbo (258); in Europe, Antwerp (268), Rotterdam (264) and Hamburg (206); in Africa, Tangier Med (137); and in Latin America, Cartagena (130). Among 12,748 direct connections, 6,017 are handled by only one shipowner (47.2%). Some 21.6% of connections are handled by two shipowners, and 31.2% of connections are handled by three or more shipowners. The ports between which cargo is transported by the highest number of shipowners are: Nigbo—Shanghai (52), Port Klang—Singapore (41), Busan—Shanghai (38), Antwerp—Rotterdam (24) and Hamburg—Rotterdam (23) (Hoffmann and Hoffmann 2020).

Figure 5.3 presents the hub ports with the highest number of feeder connections to other ports providing cargo handling services.

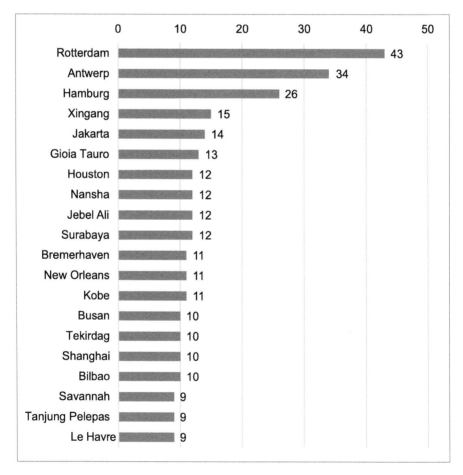

FIGURE 5.3 The ports with the highest number of feeder connections and short-sea shipping services in the world in 2020.

Source: Data from Hoffmann and Hoffmann (2020).

Exporters and importers of goods are interested in the highest accessibility to the market of maritime container shipping. The measure of accessibility is the Liner Shipping Connectivity Index (LSCI), which is regularly published by the UNCTAD for all the countries that use regular container shipping lines (Country and Port Level Liner Shipping Connectivity Index 2021). The index includes six factors (Notteboom, Pallis and Rodrigue 2022):

- the number of regular container connections per week in a particular country
- annual cargo handling capacity of ports, stated in TEU
- the number of regular container shipping lines from and to a particular country
- the number of shipowners who provide container shipping services from and to a particular country
- the average shipping capacity of container vessels (stated in TEU) that call at ports in a particular country
- the number of countries with which a particular country is connected by regular container shipping lines

The LSCI refers to most countries that have access to maritime container terminals (Table 5.6).

TABLE 5.6

The Value of the LSCI for the Selected Countries in the Years 2005, 2010, 2015, 2019 and 2020

Years Country	2005	2010	2015	2019	2020
Australia	28.0	30.8	32.5	34.2	37.2
China	108.2	143.5	138.9	151.9	162.1
Denmark	24.2	26.7	45.8	49.6	46.4
Egypt	42.2	47.5	59.0	66.7	68.5
Estonia	6.5	5.7	8.5	10.9	8.6
Finland	10.1	8.3	16.2	16.7	15.0
France	70.0	74.9	68.7	72.6	77.4
Greece	29.0	34.2	45.4	60.9	60.3
Spain	58.1	74.3	82.1	84.2	89.8
Netherlands	79.9	89.9	82.9	88.03	91.0
Hong Kong	96.7	113.6	94.2	89.4	93.6
India	36.8	41.4	49.4	55.5	57.2
Japan	66.7	67.4	74.7	71.2	87.5
Canada	39.8	42.4	39.0	42.8	47.2
South Korea	73.0	82.6	98.3	105.1	108.2
Lithuania	5.8	9.5	13.7	20.7	14.2
Germany	78.4	90.8	86.9	82.8	83.3
Poland	7.5	26.1	47.0	51.7	52.2
Russia	12.7	20.9	47.2	38.0	34.6

(Continued)

TABLE 5.6

The Value of the LSCI for the Selected Countries in the Years 2005, 2010, 2015, 2019 and 2020 (continued)

Years

Country	2005	2010	2015	2019	2020
USA	87.6	82.4	89.4	90.0	103.8
Taiwan	63.7	64.3	60.4	67.7	84.8
Great Britain	79.5	87.5	86.9	84.9	90.9
Italy	62.0	59.5	65.0	72.8	69.5

Source: Data from UNCTAD Stat (2020b).

The time when container vessels stay at ports is shorter than for other vessels, and in 2019, it was 0.69 of a day, whereas bulk carriers stayed at ports 2.05 days on average. The shortest times of handling container vessels at container terminals recorded in days were in Japan (0.35), Taiwan (0.44), Hong Kong (0.53), South Korea (0.58), China (0.60) and Spain (0.65) (UNCTAD 2020, 71). A short stay of a vessel at a port generally indicates high commercial competitiveness of that port. However, a longer stay at a particular port does not always mean that the port is less efficient because the shipowner might just decide for a longer stay in order to buy some goods or services there.

The highest numbers of containers transported by sea are handled in China (29.8% of the global turnover), the USA (6.8%), Singapore (4.7%), South Korea (3.5%) and in Malaysia (3.2%), as seen in Table 5.7. During the years 2015–2019, the volume of container handling operations increased at ports located in Greece (by 12.1%), Oman (by 12.1%), Israel (12.0%) and Poland (11.3%) (ISL Shipping Statistics and Market Review 2020, 22).

The list of ports indicating the highest volume of container handling in the world is dominated by Chinese ports. The largest container port in Europe takes the tenth position on that list. During the years 2010–2019, the highest dynamics of cargo handling operations, given in TEU, was observed at three Chinese ports, namely Xiamen, Ningbo and Guangzhou (Table 5.8).

In 2019, the largest Polish container port, Gdańsk, handled containers of the total capacity of 2.07 million TEU and held the 79th position in the world and the 16th position in Europe. During the years 2010–2019, all the ports from the top 20 list, except Hong Kong, recorded an increase in the volume of container handling operations, given in TEU.

Considering the volume of container turnover, European ports give way to numerous Asian ports; however, their importance to national economies of their countries is not less significant (Table 5.9).

During the years 2015–2019 in Europe, the highest growth rates in container handling operations were recorded at the terminals located at Gdańsk (Poland), Piraeus (Greece) and Barcelona (Spain). Vast financial investment into modernization of the existing container terminals or construction of new facilities undoubtedly contributed to that fast growth rate. Table 5.10 presents the largest container ports, according

TABLE 5.7

Top 20 Countries with the Highest Volume of Container Handling Operations at Seaports (Given in Million TEU)

No.	Country	2010	2011	2012	2013	2014	2015	2016	2017	2018	2019
1	China	132.0	146.4	159.3	174.4	185.1	193.7	197.8	222.1	233.2	242.0
2	USA	42.0	42.6	43.5	44.3	46.2	47.9	48.4	52.1	54.8	55.5
3	Singapore	29.1	30.6	32.3	33.4	34.7	31.7	31.7	33.7	37.4	38.0
4	South Korea	18.5	20.5	21.5	23.4	24.8	25.5	26.4	27.4	28.9	29.0
5	Malaysia	16.8	18.8	20.8	21.1	22.4	24.0	24.6	23.8	25	26.2
6	UAE	15.2	16.9	18.1	18.7	20.2	21.2	20.4	19.1	19.0	19.2
7	Japan	19.0	18.1	19.7	20.5	20.7	20.1	20.3	22.0	22.6	21.7
8	Spain	12.5	13.9	14.0	13.9	14.2	14.2	14.9	16.0	17.2	17.4
9	Taiwan	12.5	13.4	13.9	14.0	15.0	14.5	14.9	14.9	15.3	15.3
10	Netherlands	11.4	12.0	12.1	11.8	12.5	12.4	12.6	13.9	14.7	15.0
11	Germany	14.7	18.3	18.4	18.9	19.9	19.2	19.4	19.7	19.7	19.6
12	Vietnam	6.4	6.9	7.6	8.5	10.0	11.1	11.1	12.0	13.0	13.7
13	Belgium	11.1	10.9	10.7	10.7	11.1	11.2	11.5	12.0	12.7	13.6
14	Italy	8.0	8.7	8.6	9.5	9.7	9.4	9.8	9.9	9.9	10.0
15	Turkey	5.7	6.5	7.4	8.3	8.5	8.3	8.9	10.1	10.9	11.7
16	UK	8.2	8.1	8.0	8.2	9.5	9.8	10.2	10.2	10.3	10.3
17	India	8.9	9.9	10.0	10.6	11.3	11.9	12.1	15.4	16.9	17.1
18	Thailand	6.8	7.3	7.8	8.4	9.2	9.5	10.0	9.9	10.2	10.8
19	Australia	6.4	5.9	7.2	7.2	7.4	7.6	7.6	8.0	8.5	8.3
20	Philippines	5.6	5.3	5.6	5.8	6.2	7.2	7.4	8.1	8.7	9.0
	The world	542.8	584.2	618.2	648.9	680.5	692.4	703.5	757.1	795.7	811.2

Source: Data from UNCTAD Stat (2020a).

TABLE 5.8

The 20 Largest Container Ports in the World in 2010–2019 (in million TEU)

No.	Port	Country	2010	2017	2018	2019	Average growth 2010–2019 (%)
1.	Shanghai	China	29.0	40.2	41.9	43.3	4.9
2.	Singapore	Singapore	28.4	33.7	36.6	37.2	3.0
3.	Ningbo	China	13.1	24.5	26.7	27.5	8.6
4.	Shenzhen	China	22.3	25.0	25.8	25.5	1.5
5.	Guangzhou	China	12.5	20.1	21.5	22.7	6.9
6.	Busan	South Korea	14.2	20.0	21.7	22.0	5.0
7.	Qingdao	China	12.0	18.4	19.3	21.0	6.4
8.	Hong Kong	Hong Kong	23. 7	20.8	19.6	18.3	-2.8
9.	Tianjin	China	10.1	15.2	16.1	17.3	6.2
10.	Rotterdam	Netherlands	11.1	13.7	14.5	14.8	3.2

(Continued)

TABLE 5.8

The 20 Largest Container Ports in the World in 2010–2019 (in million TEU) (continued)

No.	Port	Country	2010	2017	2018	2019	Average growth 2010–2019 (%)
11.	Dubai	UAE	11.6	15.4	15.0	14.8	2.8
12.	Port Klang	Malaysia	8.9	12.0	12.3	13.6	4.8
13.	Antwerp	Belgium	8.5	10.5	11.1	11.9	3.8
14.	Xiamen	China	5.2	10.4	10.7	11.1	8.8
15.	Kaohsiung	Taiwan	9.2	10.3	10.4	10.4	1.4
16.	Los Angeles	USA	7.8	9.3	9.5	9.3	2.0
17.	Hamburg	Germany	7.9	8.8	8.7	9.3	1.8
18.	TanjungPelepasSingapu	Malaysia	6.3	8.3	9.0	9.1	4.1
19.	Dalian	China	5.2	10.8	9.9	8.8	5.9
20.	LaemChabang	Thailand	4.6	7.8	8.1	8.0	6.3

Source: Data from ISL Shipping Statistics and Market Review (2020, 8).

TABLE 5.9

European Ports with the Highest Volume of Container Turnover in 2015–2019 (Given in Million TEU)

No.	Port	Country	2015	2019	Average growth 2015–2019 (%)
1	Rotterdam	Netherlands	12.23	14.81	4.9
2	Antwerp	Belgium	9.65	11.86	5.3
3	Hamburg	Germany	8.82	9.26	1.2
4	Piraeus	Greece	3.36	5.65	13.9
5	Valencia	Spain	4.61	5.42	4.1
6	Algeciras	Spain	4.52	5.12	3.2
7	Bremerhaven	Germany	5.55	4.86	−3.3
8	Gioia Tauro	Italy	3.03	4.15	8.2
9	Felixstowe	Great Britain	4.04	3.56	−3.0
10	Barcelona	Spain	1.96	3.31	13.9
11	Dublin	Ireland	2.22	2.81	6.1
12	Le Havre	France	2.56	2.76	1.9
13	Marsaxlokk	Malta	3.03	2.72	−2.9
14	Genoa	Italy	2.24	2.33	0.9
15	St. Petersburg	Russia	1.71	2.22	6.7
16	Gdańsk	Poland	1.04	2.07	14.7

Source: Data from ISL Shipping Statistics and Market Review (2020, 23–24).

TABLE 5.10

The Largest Container Ports in Asia (Excluding Chinese Ports) and Both Americas (According to the Volume of Container Handling Operations in 2019, Given in Million TEU)

No.	Asia (excluding Chinese ports)	North America	South and Central America
1	Singapore 37.2	Los Angeles (USA) 9.34	Colon (Panama) 4.38
2	Busan (South Korea) 21.99	Long Beach (USA) 7.63	Santos (Brazil) 3.90
3	Port Klang (Malaysia) 13.58	New York/New Jersey (USA) 7.47	Manzanillo (Mexico) 3.07
4	Koahsiung (Taiwan) 10.43	Savannah (USA) 4.60	Cartagena (Colombia) 2.93
5	Tanjun Pelepas (Malaysia) 9.1	Seaport Alliance (USA) 3.78	Balboa (Panama) 2.90
6	LaemChabang (Thailand) 8.11	Vancouver (Canada) 3.40	Callo (Peru) 2.31
7	Tanjunh Priok (Indonesia) 7.6	Houston (USA) 2.99	Guayaquil (Equator) 1.94
8	Colombo (Sri Lanka) 7.23	Virginia (USA) 2.94	San Antonio (Chile) 1.71
9	Ho Chi Minh (Vietnam) 7.22	Oakland (USA) 2.50	Kingston (Jamaica) 1.65
10	Hai Phong (Vietnam) 5.13	Charleston (USA) 2.44	San Juan (Puerto Rico) 1.51

Source: Data from Statista (2021f, 29, 30, 31).

to the volume of container handling operations in 2019 in both Americas and in Asia (excluding Chinese ports).

Shipping containers by sea must undergo optimization stemming from the necessity of repositioning empty containers and delivering them to the depots they come from—this fact has to be taken into consideration.

5.5 EMPTY CONTAINERS IN INTERNATIONAL MARITIME TRADE (REPOSITIONING)

In a situation that is optimal for container operators, as soon as containers are delivered to their consignee and stripped, they should be loaded with cargo at the same place and shipped to another destination. However, in practice, some containers stay unused for some time. Empty containers generate serious problems in international maritime trade. The lack or an insufficient number of empty containers at a place where they are stuffed—that is, filled with cargo—inhibits international trade. As a result, carriers receive fewer containers for transportation, and this is, in turn, translated into their underutilized shipping capacity. Some disadvantageous situations may take place when at their destination place, where they are delivered and stripped, containers cannot be then stuffed with cargo within a short time, and it is necessary to store them there for some time or there is no demand for them at that place at all. In the latter case, empty containers must be sent to a place where there is a demand for them. It should be emphasized that costs of transporting empty containers are almost the same as transporting full containers. In maritime transport, container operators face the most serious economic problems whenever there is a big difference between the number of containers unloaded from vessels at a particular port and the number of containers that are loaded on vessels at that place. In such situations, empty containers must be repositioned—transported from the place where there is too many of them to a place where there is a shortage of containers.

The main reason for systematic accumulation of empty containers at ports in a particular country, at the simultaneous shortage of containers for stuffing with cargo at ports of another country, is the lack of balance in international trade in terms of the volume of transported cargo. A country that imports more than it exports faces systematic accumulation of empty containers in its territory. At the same time, countries that export more than they import, in terms of weight or volume of cargo, may face a shortage of containers. Whenever such imbalance can be observed for a longer period of time, it is necessary to reposition large numbers of containers among the partners who face a shortage or an excessive accumulation of empty containers. This is usually related to higher costs of land and maritime transport. It is estimated that 20% of the world's cargo container stock are empty containers, waiting for repositioning to a place where they are needed (Paradigma Unternehmensberatung GmbH 2021).

Disproportions in the flows of containers back and forth, including empty backhauls, can be observed along almost all maritime shipping routes (Table 5.11).

A very big difference in the capacity of repositioned containers can be observed along the Pacific route, along which 13 million TEU more were transported in 2019 from East Asia to North America than from North America to East Asia. This results from the volume and structure of trade between those two vast geographical regions that are significant to the world's economy. Large disproportions can be also observed in maritime trade between Europe and East Asia. In trans-Atlantic maritime trade, more

TABLE 5.11

Disproportions in Container Shipping along the Three Most Important Maritime Routes in 2014–2020 (in Million TEU)

	Trans-Pacific route			Asia–Europe route			Trans-Atlantic route		
Years	East Asia–North America	North America–East Asia	Difference	Europe–East Asia	East Asia—Europe	Difference	North America–Europe	Europe–North America	Difference
2014	16.2	7.0	9.2	6.3	15.5	9.2	2.8	3.9	1.1
2015	17.4	6.9	10.5	6.4	15.0	8.6	2.7	4.1	1.4
2016	18.2	7.3	10.9	6.8	15.3	8.5	2.7	4.3	1.6
2017	19.4	7.3	12.1	7.1	16.4	9.3	3.0	4.6	1.6
2018	20.8	7.4	13.6	7.0	17.3	10.3	3.1	4.9	1.8
2019	20.0	7.0	13.0	7.2	17.5	10.3	2.9	4.9	2.0
2020*	18.1	7.0	11.1	6.9	16.1	9.2	2.8	4.7	1.9

Note: * Forecast.

Source: Data from UNCTAD (2020, 15).

cargo-stuffed containers are transported from Europe to America than in the other way around. Such disproportions force shipowners to reposition empty containers.

In 1994, the UN Convention on the Customs Treatment of Pool Containers Used in International Transport was accepted (Convention on Customs Treatment of Pool Containers used in International Transport 1994). The members of the convention jointly use a pool of containers. In order to decrease transportation of empty containers and to increase the efficiency of container turnover, the parties of the convention should neither impose any economic limitations nor charge any customs duties and import taxes on containers that have been previously transported out of a particular country, and then they are transported back to that country or when the same number of containers of the same type have been transported out of a particular country and then transported back into that country. In accordance with the convention about the container pool, its parties (Paradigma Unternehmensberatung GmbH 2021):

- exchange containers among themselves in international transportation of goods;
- keep records for each type of containers to register the repositioning of containers that are exchanged in this way; and
- undertake to deliver to one another the number of containers of each type that is indispensable to compensate, over periods of 12 months, the outstanding balances of the accounts so kept, to ensure that each member of the pool can maintain the balance between the number of containers of each type that the particular member has placed at the disposal of the pool and the number of containers received to the disposal of that member in their territory.

Furthermore, the convention requires its members to implement facilitations in the field related to the turnover of container parts necessary for repair work and container accessories.

Information related to the repositioning of empty containers concludes the considerations of Part 1 dedicated to maritime containerized transport (with particular focus on the current status quo of the structure and main processes of maritime containerized transport), which allows the authors to determine specific challenges focused on a holistic approach toward costs in maritime containerized transport.

NOTES

1. Drewry Shipping Consultants Limited (Drewry)—since 1970, it has been one of the most influential research centers that offer analysis of maritime economy and global maritime shipping.
2. The CIMC Group is the world's leader in the production of maritime containers, covering over 50% of the global demand for that products. It is also the world's leader in the production of semitrailers, supplying over 150,000 semitrailers for various purposes to its customers annually. The company provides employment to over 60,000 workers in the world, and at present, it is one of the largest global employers in the sector of heavy load transport.
3. One of the world's consulting companies specializing in maritime shipping under the Shipping Intelligence Network (SIN).
4. The Official Journal of the European Union, Amendment no. 27, art. 22 and (new) P6_TA(2005)0086.

Part 2

Economic Challenges to
Maritime Containerized Transport

6 Transport and Container Handling Costs in International Maritime Transport

6.1 ALLOCATION OF TRANSPORT AND CONTAINER HANDLING COSTS TO EXPORTERS AND IMPORTERS

In international trade, goods must cover the distance between exporters and importers to be delivered to their destination places—that is, goods must be loaded onto a truck or a railway car, delivered to a port, loaded onto a vessel, transported to their destination port, unloaded, handled and finally delivered to their ultimate consignees. The right and responsibility to organize transportation and the obligation to cover costs related to that process along the particular section or the entire route are imposed on a contracting party (an exporter or an importer), whom—according to the provisions of the contract—the management of carriage (responsibility for transportation) is assigned to. This is translated into assuming a particular trade rule. In practice, management of carriage can be of a various scope; hence, it is possible to select a proper trade rule for a particular transportation process. Trade rules are commonly applied in international trade because their application streamlines the entire shipping process.

Considering container turnover, the rules listed in various sets of trade terms can be applied as such:

- Incoterms[1]
- Combiterms[2]
- RAFTD (Revised American Foreign Trade Definitions)

As mentioned prior, in practice, the scope of management of carriage can be different. While deciding about accepting responsibility for the management of carriage or leaving it to the other party, one's own economic capabilities and experience in the field of transport organization should be taken into consideration, as well as all the risk related to accepting management of carriage, a tradition followed in trade of particular goods and other factors (Marciniak-Neider and Neider, eds. 2014).

This is translated into accepting a particular trade formula. Incoterms come as the most common set of rules providing relevant formulas. They are based on a division of the entire transport process (including container transportation as well) between the seller (the exporter) and the buyer (the importer) into the following stages (BBA Transport System 2021):

- receiving the consignment (a container) from the manufacturer
- transporting it to a warehouse/port/terminal/airport (with the use of a complementary means of transport)
- storing and other terminal operations
- loading it onto the main means of transport
- transporting the consignment by the main mode of transport
- unloading the consignment from the main means of transport
- storing and other terminal operations
- transporting the consignment to its consignee (with the use of a complementary means of transport)
- receiving the consignment (a container) by its consignee

Incoterms 2020 are divided into four groups (C, D, E, F). The rules are divided by fees incurred, risk, responsibility for formalities and issues related to import and export. The C group (Main Carriage Paid)—the CFR, CIF, CPT and CIP formulas—indicates that the seller enters into a shipping contract with a forwarder and incurs the costs. Hence, the seller (the exporter) is held responsible for carrying out the export freight clearance, and at the moment of dispatching the consignment, the risk is transferred onto the buyer (the importer). It is assumed that after the loading process is finished, all the issues resulting from the shipping process become the responsibility of the buyer. The D group (Arrival) includes the DAP, DPU and DDP, which impose an obligation to deliver the goods to a particular destination place (or to a destination port) on the seller (the exporter). The E group includes only one rule, namely Incoterms EXW, according to which the seller makes the goods available to the buyer (the importer) at the premises defined by the seller (the exporter). The seller is not responsible for export freight and customs clearance. Also, the seller does not accept the risk and does not incur costs related to loading operations. The F group (Main Carriage Unpaid) includes the FCA, FAS and FOB, which impose the obligation to carry out the export freight clearance on the seller. The costs of transport and insurance are incurred by the buyer (the importer). Some rules are applied in all modes of transport: EXW (Ex-Works, at the named place), FCA—Free Carrier (at the named place), CPT—Carriage Paid To (the named place of destination), CIP—Carriage and Insurance Paid To (the named place of destination), DAP—Delivered at Place (the named place of destination), DPU—Delivered at Place Unloaded (the named place of destination) and DDP—Delivered Duty Paid (the named place of destination). Other rules are dedicated to maritime transport. These are FAS, FOB, CFR and CIF, and they are applied as follows:

- The Incoterms 2020 FAS rule (Free Alongside Ship, named port of shipment) states that responsibility for transportation and the risk of the seller end at the moment of delivering the goods alongside the vessel at a designated port and they do not cover the loading of the goods onto the vessels. From that moment, the entire risk and all the costs related to the goods are transferred to the buyer.

- The Incoterms 2020 FOB rule (Free On Board, named port of shipment) states that the moment when the goods are loaded onto the vessel is the moment when the risk and costs related to the goods are transferred from the seller to the buyer. The seller incurs the cost of loading the goods onto the vessel.
- The Incoterms 2020 CFR rule (Cost and Freight, named port of destination) states that the seller is not responsible for incurring the insurance costs (from the moment when the goods are loaded at the port from where the goods have been dispatched). The buyer is responsible for covering insurance costs. The delivery is considered as done at the moment when the goods are loaded onto a vessel of the shipowner indicated by the seller.
- The Incoterms 2020 CIF rule (Cost Insurance and Freight) holds the seller responsible for paying and entering into the insurance contract in favor of the buyer. The seller also incurs the freight costs.

Combiterms and RAFTD contain trade rules referring to a more specific division of liabilities and obligations of the seller and the buyer. They also specifically define costs incurred by the transaction parties (E-logistyka 2021). Combiterms supplement some rules with additional information on costs (rules 001–023), which is not possible in the case of Incoterms, considering their postulated universality. In practice, a mixed formula can be applied: Incoterms to specify the particular rule and Combiterms to divide costs. For example, in the 007 CIP rule (Carriage and Insurance Paid), the seller incurs the same costs as those specified in the 006 CPT rule and, additionally, is responsible for transit documents. The buyer is responsible for costs incurred starting from the seller's country borders. The 008 CFR rule (Cost and Freight, applying to maritime transport), apart from responsibilities similar to those specified in the 007 CIP rule, states that the seller is also responsible for transport costs that have to be incurred starting from the seller's country land borders to the buyer's country land borders. The buyer covers transport costs starting from the buyer's country borders. In 009 CIF (Insurance and Freight), the seller's responsibilities are similar to those specified in the CFR rule; however, the seller additionally incurs costs related to delivering the goods to the land border of their destination country. Other costs are incurred by the buyer. The 013 CIP rule (Carriage and Insurance Paid) states that management of carriage is imposed on the seller, up to the particular destination place. The seller is responsible for organizing and incurring transport costs along the entire delivery route to the consignee's warehouse, including import customs clearance, documents and freight fees.[3] As the previous examples indicate, the Combiterms rules allow the parties to allocate costs in a precise way.

The Revised American Foreign Trade Definitions (RAFTD) 1941 (Kasprzyk 2019) are commonly applied in the USA, Mexico and the countries of Middle America. Undoubtedly, it would be advisable to discuss the analogy and similarities in the meaning and functionalities to the previously mentioned Incoterms. It should be also emphasized that their interpretation can be often inconsistent. This fact implicates some (and sometimes even fundamental) discrepancies in the application of the apparently the same trade rules. Hence, while referring to the particular abbreviations, it should be clearly indicated which interpretation the parties mean. The terms applied in the RAFTD 1941 include the Ex rule, six FOB rules (A-FOB, B-FOB, C-FOB, D-FOB,

E-FOB, F-FOB), the FAS, CF, CIF rules and the Ex Dock rule. In comparison to the Incoterms rules, the obvious differences that can be easily pointed out are six variations of the main FOB rule (Free on Board), whereas the European version (Incoterms) defines only one FOB rule.[4] Over the recent years, however, the RAFTD rules have been gradually replaced by the Incoterms rules (Kasprzyk 2019),[5] although they are still commonly applied in the USA, Mexico and the countries of Middle America.

Other significant components of maritime transport costs incurred in container turnover are port fees.

6.2   PORT FEES THAT ARE CHARGED ON MARITIME CONTAINER OPERATORS FOR THE USE OF PORT FACILITIES AND CONTAINER TERMINALS (THE EUROPEAN APPROACH)

Port authorities charge container vessels and other vessels calling at the port with fees for using the port infrastructure. To calculate port fees correctly, captains of vessels calling at a particular port or their agents must electronically notify the port chief dispatcher about an arrival or a departure of the vessel to or from the port. Usually, shipowners of container vessels are charged with two types of fees (Port of Gdynia Authority SA.2021):

- The tonnage fee—charged for one arrival to and departure from the port, a transit through the port area, collection of ship waste: waste oil, solid waste and sewage. The amounts of port fees are calculated according to the gross tonnage (GT) of vessels, their valid International Tonnage Certificate or a valid Ship Safety Certificate. Tonnage fees include collection of waste (waste, waste oil, sewage) up to the limit established in m³.
- The wharfage fee—charged for the use of a wharf or the port harbor. The amounts of the wharfage fees are calculated according to the vessel types, vessel sizes—that is vessel capacity stated in GT—and the period of time when a vessel occupies the berth at the port wharf.

Fees charged at ports are included in the tariffs of port fees that are publicly available. Fees charged at ports are varied and their amounts affect the competitiveness of a particular port in the market of maritime containerized transport. The following is a tariff of tonnage and wharfage fees charged by the Port of Gdynia, Poland, presented as an example (Table 6.1).

It should be emphasized that port fees charged on container vessels are lower than those charged on general cargo vessels or bulk carriers. There is a discount on the tariff rates, and its amount depends on the frequency of calls at the Port of Gdynia, Poland, made by vessels of a particular shipowner. The discount amounts on the tonnage fees calculated for regular line vessels that call at the Port of Gdynia, Poland, vary between 60% (for vessels calling at least 12 times a week), 25% (for vessels calling once a week) and 10 % (vessels calling not less rarely than once a quarter) (Port of Gdynia Authority SA.2021,6). The discount is usually calculated collectively for all the vessels operated under a particular regular line and under one trademark. The amounts of wharfage fees for seagoing vessels at Polish ports are presented in Table 6.2 (as an example).

TABLE 6.1
Tonnage Fees for Seagoing Vessels Charged by the Port of Gdynia, Poland (since 1st August 2020)

No.	Type and size of vessels	Amount of fees (PLN/1GT)
1.	Car carriers	0.65
2.	General cargo vessels	2.03
3.	Refrigerated vessels	2.03
4.	Container vessels	0.78
5.	Ro-ro vessels	0.95
6.	Bulk carriers	2.15
7.	Passenger vessels	0.39
8.	Ferries	0.56
9.	Tankers	2.48
10.	Tugs, pusher tugs, push trains and towing trains	1.55
11.	Cutters and fishing boats less than 35m	0.00
12.	Other seagoing vessels	2.03

Source: Data from Port of Gdynia Authority SA (2021, 6) (accessed: 20th March 2021).

TABLE 6.2
Wharfage Fees for Seagoing Vessels at Polish Ports (PLN/GT)

No.	Type and size of vessels	Time of using the port infrastructure	Amounts of fees (PLN/1 GT)
1.	Ferries, ro-ro ships, car carriers, passenger vessels	For the time of using the wharf for operational or commercial purposes and for the first 4 hours after that	0.18
2.	Other vessels	For the time of using the wharf for operational or commercial purposes and for the first 4 hours after that	0.48
3.	Each vessel	For each commenced 4-hour period after 4 hours passed from the time of completion of operational or commercial operations	0.18

Source: Data from Port of Gdynia Authority SA (2021, 6) (accessed: 20th March 2021).

Apart from the previous fees, container shipowners are charged for specialist services provided by commercial companies, such as towage, moorage, cargo expertise and inspection, pilotage, securing containers on board against moving during the sailing, cleaning vessel holds and other services (Port of Gdynia and Maritime Services 2009).[6]

Tonnage and wharfage fees come only as fragments of a larger picture showing problems related to costs of container shipping and handling operations that must be incurred by customers in international maritime trade. Another significant cost group in the processes of container shipping and handling that must be considered by customers in international maritime trade includes costs related to customs (customs duties and taxes) clearance.

6.3 FEES CHARGED FOR CUSTOMS BROKERAGE SERVICES UNDER THE CUSTOMS PROCEDURES FOLLOWED IN CONTAINERIZED CARGO SHIPPING (THE EUROPEAN APPROACH)

Shipping containers by trucks, where it is necessary to cross customs borders, including feeder transportation of containers from seaports (terminals), were significantly facilitated by the Convention on International Transport of Goods Under Cover of TIR Carnets (the TIR—Transport International Routier—Convention) of 1959, which was updated in 1975 (TIR Convention 2019). In the provisions of the convention, a container is considered to be a packaging unit dedicated to the purposes of international transportation of goods under the customs bond. It means that there are neither customs duties collected at the border customs posts/offices nor security payments or guarantees required against any possible duties or taxes related to the import of goods into a particular customs territory. TIR carnets come as the only sureties here. Furthermore, customs bonds that are set up at the entry/exit customs posts are accepted by the customs authorities of the other countries, so trucks pass through border crossing checkpoints without opening containers and inspecting their cargo.

The TIR carnet comes both as a customs declaration and a surety (the guaranteed amount is usually EUR 100,000).[7] Transportation of goods under the TIR procedure can take place only when the shipping is started or is supposed to end outside the territory of the European Union or when the goods are transported from one place to another location in the area of the European Union in transit through the territory of a third country. Under the TIR carnet procedure, it is possible to transport all goods, except for cigarettes and liquors. The cover of a TIR carnet is presented in Annex 8.2. Depending on the number of borders to be crossed, there are several types of TIR carnets (ShipHub 2021b):

- 4-volet (voucher) carnets (for passing between two customs territories)[8]
- 4-volet-PILOT carnets that can be applied when shipping starts in any EU country and ends in a third country directly adjacent to the EU borders— e.g., Poland–Ukraine and in the opposite direction
- 6-volet carnets (up to three customs borders)

- 6-volet-PILOT carnets that can be applied when goods are transported from or to an EU country or from a third country to a third country in transit through the territory of any EU country (e.g., Switzerland–Ukraine, Turkey–Ukraine, etc., through the EU customs territory)
- 14-volet carnets (up to seven customs borders at the maximum)
- 20-volet carnets

Goods transported in containers into the EU customs territory and goods transported out of the EU are subject to a customs declaration. After that, they are assigned with one type of the following customs-approved treatment (Polish Coms Department 2021b):

1. Placing goods under the customs procedure:

   - admission to trading
   - transit
   - customs bonded warehouse of active refining/inward processing
   - processing under the customs supervision
   - temporary (interim) customs clearance[9]
   - passive refining/outward processing
   - exporting

2. Bringing the goods into the common customs territory or to a bonded warehouse
3. Sending goods back outside the EU customs territory
4. Relinquishing goods to the state

Most often, goods transported in containers undergo customs clearance procedures to be admitted to domestic or foreign trade (import or export). To obtain such a permission, it is necessary to submit a Single Administrative Document (SAD)[10] and other documents, such as an invoice (it must include information about currency and delivery terms), a bill of lading, a packing list,[11] EU conformity certificates (*fr. Conformité européenne*) and a certificate of the origin of goods if the trade involves countries classified for preferential customs duty rates.

Usually, the customs clearance procedure is carried out by a customs agency on behalf of an exporter or an importer. The export customs clearance may take place at the customs office in the exporter's area or at the border customs office. Whenever goods are exported in fully loaded containers, most often, the customs clearance procedures take place at the premises where the goods are loaded because customs clearance procedures at the port premises would mean additional costs related to customs inspections (unloading goods from a container and loading them back again, involvement of a cargo examination and commodity expertise services). If the customs clearance procedures take place at the port, two other documents are required, namely a DSK[12]/ATB[13] summary declaration of a container and a goods specification chart for LCL shipments (less than container load). Both documents are printed from the electronic system of the port terminal. If containerized goods are sent outside the EU customs territory, the ECS (Export Control System) is applied to handle digital customs declarations. After the goods have been packed into containers and after the customs clearance procedure has been completed, the container doors are secured with customs seals.

The information about the completion of the customs clearance procedures, the identification number and the seal numbers are provided in the container shipping document. The export customs office releases the goods for export and sends an Export Accompanying Document (EAD) to the submitting party. At the same time, the customs office sends electronically the respective message to the border customs office of exit to inform in advance about the container consignment that is going to appear at the border crossing point. The EAD document physically accompanies the consignment to the border crossing point, where the identification number is read with the use of a dedicated reader and then it is entered to the ECS system. After the decision is made to release the goods to the area outside the EU customs territory, the customs office of exit sends the message to the export customs office to inform about the "results of inspection at the customs office of exit," and the submitting party is provided by the export customs office with the "confirmation of the exit of goods" that replaces the 2SAD paper document and comes as a confirmation of the goods being exported, to be submitted for the VAT purposes (Neider 2012, 307–308).

The customs clearance procedures for containerized goods can take place at the EU border ports (admission to trade procedures) or at the customs offices located in the interior of the country, in the vicinity of the goods consignee's headquarters or at the premises of the importer's plant (transit procedures). In the first case, the forwarder (the customs agent) prepares a SAD document and sends it to the customs office, registering this operation in the AIS/IMPORT system. When the goods undergo an obligatory inspection, the submission of the customs declaration may take place after the required laboratory tests are completed. The customs office may resign from the opening of a container; still, they can demand to see the goods packed inside, which involves additional costs incurred by the importer. Before containers are released, the importer must duly cover all the customs duties and taxes required or to submit financial security against the customs debt. In the latter case, the forwarder declares the goods to the customs office, using the NCTS (New Computerized Transit System) electronic system, applying for handing containers over to the import Customs Office (the T1, T2 or TIR transit procedures). It is required to submit a financial security against the customs duty and tax. The ultimate customs clearance procedures take place at the import customs office. The consignee must submit the invoice, the specification of goods and other required documents. After the customs clearance is finished, the entire customs procedures are completed, and this fact is registered in the NCTS (Neider 2012, 313–314).

The entity acting on behalf of the involved parties are customs agencies that also operate at ports and terminals. The fees charged for services provided by customs agencies in the field of importing or exporting goods depend on numerous factors. The price and the time of customs clearance procedures may be affected by, among others, issues such as the place of the customs clearance, the type of goods or the mode of transport (maritime, land [trucks and railway], postal or courier consignments).

The services most often provided by customs agencies are the following (WER-SAD 2021):

- EORI (Economic Operators Registration and Identification)[14] registration
- electronic application for the Admission to Trade procedures in the AIS system/SAD document
- electronic application for the export procedure in the AES[15] system
- issuance of certificates of origin, EUR.1, ATR
- issuance of T1 and T2
- electronic completion and application for TIR carnets
- organization of inspections performed by the market surveillance authorities (WIJHARS,[16] SANEPID,[17] PIORIN)[18]
- participation in customs inspections
- issuance of temporary admission declarations

The amounts of fees charged for services provided by customs agencies (Poland as an example) range between PLN 150–200 per one customs document. Completing and legalizing EUR and ATR certificates may cost PLN 50–100. The price of issuing T1 or T2 depends mainly on the type and value of goods. The more expensive the goods and the higher the customs rates, the higher the amounts of the guarantees. Furthermore, some goods involve higher risk (e.g., excise goods) and they require individual arrangements. However, standard prices of such services range from PLN 100 to PLN 200 per submission.

For some types of goods, it is indispensable to carry out inspections. Inspections must be performed by relevant bodies (WIJHARS, SANEPID, PIORIN) before the goods are admitted to trade. Most frequently, it refers to agricultural products, food or goods in contact with food. In order to meet such requirements, a customs agency may, on behalf of the ordering party, apply for relevant inspections to respective inspectorates. Moreover, a customs agent may participate in an inspection on behalf of the importer, whose presence is not necessary in such a case. The cost of such a service is PLN 50–100 for an application and the customs agency must have an additional authorization to provide it (WER-SAD 2021).

Admission of goods to trade in the EU territory[19] is possible only after all the payments related to import of goods from outside the EU are duly made. The amount of customs duties that have to be paid for imported goods includes the customs value and the customs rate based on the customs tariff and the exchange rate. The customs value is the price declared by the importer, which is a payable price or the price that has been eventually paid and declared on the import invoice. The customs value also includes costs of transport and insurance incurred by the importer and calculated up to the EU customs border. The import of goods also undergoes VAT taxation. While importing goods from the outside of the EU territory, the entity obligated to pay the VAT is the taxpayer who is also responsible for paying the customs duties (see more in Poradnik Przedsiębiorcy 2021).

Other elements of costs related to transport and container turnover in international maritime transport are freight fees.

6.4   FREIGHT RATES IN MARITIME CONTAINER TRANSPORT

The remuneration/payment obtained by shipowners for shipping loaded or empty containers from one port to another and also for transporting bulk cargo is a sea freight.[20] The amount of the sea freight depends on many factors: types of containers (general-purpose or special-purpose containers), distance to be covered during the transportation, the level of competitiveness in the freight market and other elements. The payment for shipping is based on the sea freight tariff, which specifies freight rates for shipping goods/containers along the particular shipping routes. Usually, sea freight rates are stated in USD or EUR. In the freight market, there are also individual tariffs specified by shipowners and conference tariffs, which are freight rates uniformly charged by members of a shipping conference, who operate along the particular shipping routes or in the particular maritime areas. The amounts of freight rates are highly affected by the situation on the container shipping market (the market of shippers or carriers, the level of market monopolization, costs incurred by shipowners in relation to container shipping, competition from alternative modes of transport and random factors, such as, for example, the Covid-19 pandemic), which can disturb the regular functioning of the freight market, particularly in a short period of time.

Freight rates in container shipping are of a complex structure because there are numerous technical and organizational factors taken into consideration, which affect the providing of shipping services. Apart from port-to-port rates that define the price of maritime container shipping, starting from the cargo loading port to the cargo unloading port, there are also rates for direct point-to-point shipping, from the place where the cargo is received for transportation at the hinterland of the cargo loading port to the place where the cargo is released to its consignee, outside the cargo unloading port in the destination country. Direct rates include also point-to-port rates and port-to-point rates (Marciniak-Neider and Neider, eds. 2014, 262).

Depending on whether the shipping refers to full container load consignments or containerized general cargo, FCL-type rates (full container load) and LCL-type rates (less than container load) are applied. They differ in the way of calculating the amount charged for the unit of goods intended for transportation. In the first case, the calculation is based on a container, whereas in the latter case, the calculation refers to the weight or the capacity of the cargo. Whenever full containers are transported, the basic rate may include only a fee for shipping by seagoing vessels to a hub port, to which another fee is added for shipping by feeder vessels (TAD—Transport Additional) to the destination port, where containers are unloaded. The LCL and FCL rates are usually accompanied by a number of symbols (Marciniak-Neider and Neider, eds. 2014, 263):

- The CY/CY or Y/Y symbols refer to containers into which the cargo has been stuffed by the shipper outside the premises of the carrier and delivered to the container yard (CY) at the carrier's terminal; however, the goods are unloaded outside the carrier's facilities and at the cost and risk incurred by the consignee of the cargo.

- The CF/CES or Y/S symbols refer to containers with the cargo stuffed by the shipper outside the carrier's facilities and delivered to the container yard at the carrier's terminal; the cargo is unloaded by the carrier, at the carrier's container freight station (CFS).
- The CFS/CFS or S/S symbols refer to a delivery of the cargo to the carrier's container freight station, where the carrier loads the cargo into containers and unloads it at the carrier's container freight station, at the destination port, at the cost and risk incurred by the consignee of the cargo.
- The CFS/CY or S/Y refer to a delivery of the cargo to the carrier's container freight station, where the carrier loads it into containers that are received by the consignee at the container yard at the carrier's container terminal; the cargo is unloaded outside the carrier's facility, and all costs are to be covered by the consignee of the cargo.
- The DOOR or D or D/D symbols refer to land transport of the cargo provided by the carrier from the shipper's facilities to the loading port and from the destination port to the consignee's facilities.

The freight tariff is a list of prices for shipping services, along with the conditions and the scope of their application. The structure of the freight tariff is based on the principles applied to calculate freight rates, cargo description, a list of rates referring to the particular groups of goods or to the particular cargo. Tariffs may also include prices charged for feeder transport, land–sea transport or ship–ship transport. The most commonly applied tariffs in container shipping are the following (Kujawa 2020, 114):

- Commodity box rates (CBR)—a tariff that refers to the rates charged for fully containerized cargo (FCL—full container load); it is calculated based on the type and the value of the goods.
- Freight all kinds (FAK)—a tariff that is based on the principle referring to uniform rates for all containers; it is calculated based on a uniform (the same) rate for a container, no matter what types of cargo are transported (the calculation is based on the unit cost, enlarged by the carrier's profit margin).

The CBR tariff is applied to the cargo received from the booking parties who order transportation of containers (FCL). The FAK tariff is applied to the cargo booked by maritime and land forwarders who are multimodal or intermodal transport operators. There is also a possibility to negotiate prices for transportation services with the shipping lines, but it usually means that service contracts must be entered between carriers and booking parties. The terms of a service contract provide more flexibility in the cooperation between the contract parties; however, the booking party is obligated to deliver more cargo for transportation. Another type of prices is an *ad hoc* negotiated rate. Such rates largely depend on market conditions, and they can be lower or higher than the current CBR rates (depending on the economic conditions).

The characteristic feature of the freight system in containerized regular shipping is charging additional fees (freight additionals) by carriers. These fees are referred to as freight additional charges. They can be stated in an amount or percentage, and they can be applied periodically or permanently. The most common freight additional charges are the following (cf. Urbanyi-Popiołek, ed. 2012, 69–70):

- BAF—bunker adjustment factor[21]—an amount charged for a TEU (a fee charged by the carrier in case of an unexpected increase in fuel costs; it can be also referred to as bunker charge and FAF—fuel adjustment factor).
- CAF—currency adjustment factor—takes form of a percentage index calculated on the total freight amount or on the basic value of the freight rate referred to a container unit (applied to compensate the fluctuation of exchange rates to maritime carriers).
- ERS and EIS[22]—equipment repositioning surcharge and equipment imbalance surcharge—amounts charged for a freight unit, namely for a container of any type and any size (they might be of incidental and periodic nature).
- SSC—security (sur)charge—an additional fee resulting from the ISPS regulation for the security system; it is an amount charged for one loading unit.
- Customs documentation and export formalities fee—a documentation fee referring to the costs of customs formalities in export.
- AMS—advanced manifest surcharge—an additional fee for the prior freight manifest (it generally refers to the use of the automated communication system and processing the data on freight manifests and cost internalization).
- War risk premium—an additional fee for war risk (an extra fee to compensate the war risk in the territories and water areas of the countries along the cargo shipping route).
- C/S or PCS—congestion surcharge—an additional fee charged by the carrier to compensate the risk of delay (running out of schedule) at ports (it refers to demurrage and detention of loading units).
- DDC—destination delivery charge—a fee charged depending on the size of containers to cover costs related to crane and gantry crane operations, maneuvering containers around the container terminal, including a gate fee (a tipping fee), which is a fee charged for the particular amount of waste generated during the operations taking place at the terminal.
- Diversion charge—an additional fee charged for any changes made to the shipping route or to the initially designated destination place.
- PSS—peak season surcharge—a seasonal additional fee charged when other shipping lines are particularly overloaded with orders and there is heavy traffic generated by an increase in demand for imported goods.
- CSC—container service charge—a fee charged for the positioning of containers at the terminal.
- Heavy weight surcharge, heavy-lift charge—a fee charged for weight and heavy-lift cargo that cannot be handled with the use of standard cargo handling facilities at the terminal.
- Wharfage—a fee charged for using the wharves and wharf facilities.
- THC—terminal handling charge—a fee related to operations performed during cargo loading and unloading.[23]

The amount of rates can be also affected by a solution that has been adapted in reference to the way of receiving the cargo in a container. If the carrier receives a container stuffed by the consigner of the cargo and releases the container in the same form to the consignee, the rate is lower than in a situation when a container is stuffed by the carrier at the carrier's own cost and risk (cf. Urbanyi-Popiołek, ed. 2012, 71–72). Then the amount of the rate is higher or a relevant additional fee is applied (Polish Customs Department 2021a).

At the same time, apart from fees charged additionally to freight rates, there are discounts available in the market too. After meeting some previously agreed conditions, shippers may obtain discounts after a longer period of time. There are quantity discounts that shippers may obtain after they exceed a particular number of consignments ordered for transportation. There are also loyalty discounts for customers who use services provided by a particular carrier for a particular period of time or almost all the time.

Knowing the freight rate with all the additional fees and discounts, it is possible to calculate the amount of the freight—that is, the remuneration of the shipowner for transporting the particular lots of cargo from the port of loading to the destination port.

6.5 FEES CHARGED ON EXPORTERS AND IMPORTERS OF CONTAINERIZED CARGO FOR SERVICES PROVIDED BY CONTAINER TERMINALS (THE EUROPEAN APPROACH)

Goods exported abroad and imported from foreign countries come and go through container terminals. In export, an exporter or a forwarder delivers stuffed containers to a port yard, providing the number of a vessel and the number of the sailing. These data allow the terminal operator to put the containers on the proper storage places and to verify the data provided by the forwarder with the loading (cargo) list delivered by the shipowner's agent. A port yard is the place where containers with imported goods transported by vessels are received—it is done in accordance with the shipowner's freight manifest. The allocation of cargo handling costs to the particular consignee/forwarder takes place as a result of exchanging the bill of lading for a summary declaration and indicating the particular forwarder by the shipowner's agent to be the container consignee at a particular port.

In accordance with the THC (Terminal Handling Charge) system, the shipowner settles the costs related to the stowage and cargo handling of containers in relation to vessel–port yard means of land transport or the other way around, which account for the fundamental part of port costs, with the terminal operators, and then charges the cargo consignee with those costs arbitrarily. Other port services provided at container terminals, such as maneuvering and storage of containers, are settled by forwarders directly with the terminal operators. They also order the loading and unloading of containers onto trucks or railway cars.

Fees charged for services provided at the container terminal are published in a tariff that is often of an informational character only, underlying a negotiation ground for ordering parties. Customers who trade large quantities of cargo and who use services provided by a particular terminal for a long time can count for high discounts and preferential terms of payment. In order to attract new customers, a welcoming bonus, usually of 5%, may be granted while the first contract with an ordering party is signed. It may also be a longer free storage time of containers. Fees charged for loading, cargo handling and stowing dangerous goods are usually higher—for example, by 100% at the Baltic Container Terminal (BCT), Poland. Similarly, fees for operations involving oversized cargo are also higher. Higher fees are charged for services ordered and provided on Saturdays, Sundays and holidays. Generally, prices of services provided in the field of maritime stowage range (Poland as an example) from PLN 200 to PLN 400 per container, ship-shore-ship cargo handling operations cost from PLN 100 to PLN 200 per container, plugging a reefer container (one hour of cooling) is about PLN 20, transshipment operations cost from PLN 300 to PLN 600 and weighting one container costs PLN 90–100. The tariff of the BCT is presented in Table 6.3.

TABLE 6.3

The Tariff of Fees Charged for Services Provided at the Baltic Container Terminal, Gdynia, Poland, Stated in PLN (2021)

Containers 30′ and 45′ containers are counted as 40′ containers	Measurement unit	Full		Empty	
		20ft	40ft	20ft	40ft
Stowage	pcs	330.00	400.00	210.00	270.00
Lashing or unlashing containers on a vessel	pcs	31.00	31.00	31.00	31.00
Ship-to-shore or shore-to-ship cargo handling operations	pcs	150.00	190.00	110.00	140.00
Yard-truck or truck-yard cargo handling operations	pcs	150.00	190.00	110.00	140.00
ISPS security charge	pcs	12.00	12.00	–	–
Maneuvering	pcs	270.00	350.00	270.00	350.00
Storage—7 days included into the cargo handling rate					
from 8 to 14 days	pcs/day	11.00	22.00	11.00	22.00
from 15 to 21 days	pcs/day	22.00	44.00	22.00	44.00
from 22 to 28 days	pcs/day	28.00	55.00	28.00	55.00
from 29 to 90 days	pcs/day	85.00	150.00	85.00	150.00
from 91 days	pcs/day	170.00	300.00	170.00	300.00
Reefer containers					
Cooling (plugging and unplugging, power, monitoring)	Hours	20.00	20.00	–	–
Other operations					

TABLE 6.3
The Tariff of Fees Charged for Services Provided at the Baltic Container Terminal, Gdynia, Poland, Stated in PLN (2021)

Service	Unit				
Transshipment	pcs	470.00	610.00	320.00	410.00
Additional yard–yard, railway car–railway car, truck–truck, inside truck or insider railway car maneuvers of (full/empty) containers ordered by customers	pcs	150.00			
Changing information in the MAINSAIL system	pcs	60.00			
The retaining of goods in export order	pcs	90.00			
Weighing cargo on the truck	pcs	95.00			
Weighing cargo on the BCT tractor	pcs	95.00 + manipulation costs			
Putting on/taking on a tarpaulin on a container	pcs	80.00			
Sticking on or removing IMO labels	container	120.00			
Stacking with or without lashing (platforms, bolsters, rolling stock)	1 set	160.00			
Using emergency folding trays		450.00 for each commenced day, longer than 24 hours			
Waiting of the loading team on a container vessel	Hour	1,550.00			
Assistance of an STS gantry crane	30 min	2,020.00			
Taking off or putting on a cover on the vessel hold opening	pcs	350.00			
Re-stowage in one hold	pcs	150.00			
Ship-shore-ship re-stowage	pcs	380.00			
Railway operations					
Train inspection	Railway car	15.50			
Preparation of a railway car for loading	Railway car	16.00			
Handling of the INCOS electronic platform by a BCT employee	Railway car	15.00			
Documentation brokerage	Train	60.00			
Additional fees for delayed notification of containers delivered by railway	pcs	150% of the rate charged for container handling			

(Continued)

TABLE 6.3

The Tariff of Fees Charged for Services Provided at the Baltic Container Terminal, Gdynia, Poland, Stated in PLN (2021) (continued)

Fixing lashings on railway cars with the use of additional security measures	Pcs	40.00
Stay of a train at the terminal	Train/hour	350.00
Waiting of the railway loading team	hour	1100.00

Source: Data from BCT (2021) (accessed: 10th October 2021).

The total payment for shipping containers by sea can be affected by the previously mentioned fees for storage, demurrage and detention of containers related to terminal operations. Storage fees include fees for storing containers at the terminal, starting from the moment they have been delivered to the terminal yard to the moment they are taken away. These costs are incurred directly by the forwarder or the shipowner who stores the containers at the terminal. Usually, there is some free-of-charge time period of storing, after which storage fees are charged per TEU/day. Considering maritime terminals, the storage fees are applied for export and import containers, and they can be proportionally increased for longer periods of storage (Waldmann 2016, 189). Table 6.4 presents an example how storage fees are calculated by the MSC shipowner at the Polish ports.

Demurrage fees are charged in a situation when the time from the moment of unloading full containers to the moment of their removal from the port yard is longer (in days) than the free-of-charge storage time. The amounts of demurrage fees are calculated as the mathematical product of demurrage rates and the number of demurrage days, with the consideration of days when those fees are not charged. Hence, the demurrage fees are calculated for keeping (the cargo) containers for the shipowner for a period of time that is too long, as opposed to storage. The demurrage fee is charged on export and import containers; however, the rates are usually lower for export containers than for import containers. In export, when containers are not taken and loaded on a vessel that is going to transport them to their destination place specified in the relevant documents (e.g., because of a system failure on the forwarder's side or there is a delay in packing containers by the forwarder until the cutoff time is up), both storage and demurrage fees can be charged. If containers are still on the yard because of the shipowner's fault, the shipowner is charged with the storage fees; however, as the owner of the containers, the shipowner is not charged with any demurrage fees. Table 6.5 presents an example of calculating demurrage fees by the MSC shipowner at the Polish ports.

Detention fees are charged starting from the day when full containers are taken from the port to the day when empty containers are sent back to the port (container terminal) or to a depot specified by the shipowner. The aim of detention fees charged for detention and keeping containers on the yard for too long is to accelerate their

TABLE 6.4

Storage Costs (Import/Export) Incurred by the MSC Shipowner at the Polish Ports (EUR/TEU/DAY)

Import storage costs		Export storage costs	
One day of storage	Daily cost for 20′ container	One day of storage	Daily cost for 20′ container
0–10	Free of charge	0–9	Free of charge
11–14	EUR 4	10–14	EUR 4
15–30	EUR 5	15–30	EUR 5
31–60	EUR 10	31–60	EUR 10
Over 60 days	EUR 20	Over 60 days	EUR 20

Source: Data from Maersk (2021a), MSC (2022) (accessed: 22nd March 2021).

TABLE 6.5

The (Import/Export) Demurrage Costs of the MSC Shipowner at the Port of Gdańsk (EUR/TEU/Day)

Maersk Gdańsk import demurrage			Maersk Gdańsk export demurrage		
Day of storage	20′ container	40′ container (and high cube)	Day of storage	20′ container	40′ container (and high cube)
0–12	Free of charge	Free of charge	0–9	Free of charge	Free of charge
13–21	EUR 33	EUR 43	10–17	EUR 13	EUR 13
Over 21	EUR 53	EUR 73	Over 18	EUR 23	EUR 23

Source: Data from Maersk (2021a), MSC (2022) (accessed: 22nd March 2021).

return and to prevent using them as cargo storage containers. Hence, if full containers delivered to the port are supposed to stay there for a longer period of time, then it is more economical to keep containers in a port warehouse.[24] The fees are charged in import and export and their amounts and the time when the charging starts depend on "time free of charge" and on the means of transport (when import containers are transported by road transport, there are three additional days free of charge; if they are transported by railway, there are eight days free of charge). Considering trends to simplify calculation of fees and tariffs, most shipowners implement a specific way of cost calculation, combining demurrage and detention fees into one integrated fee—CDD (combined demurrage detention). The CDD is charged for using containers from the moment they are unloaded from a vessel to the terminal to the moment when empty containers are repositioned to one of the shipowner's depots (without splitting costs in terms of the place they are generated at—the terminal or outside the terminal). In a way, this method forces operators to release empty containers as soon as possible and to return them to the shipowner's depots within the shortest possible time after the service of maritime shipping has been completed. Table 6.6 presents an

TABLE 6.6

Combined Demurrage Detention Fees of the MSC Shipowner at the Polish Ports of Gdańsk and Gdynia (EUR/Container Type/Day)

	20′ container	40′ container	40′ and high cube containers
Gdynia CDD fees			
Day of storage		Daily cost	
0–7	Free of charge	Free of charge	Free of charge
8–14	EUR 15	EUR 30	EUR 30
Over 14 days	EUR 40	EUR 80	EUR 80
Gdańsk CDD fees			
	20′ container	40′ container	40′ and high cube containers
Day of storage		Daily cost	
0–14	Free of charge	Free of charge	Free of charge
15–21	EUR 15	EUR 15	EUR 15
Over 21 days	EUR 40	EUR 40	EUR 40

Source: Data from Maersk (2021a, 2021b).

example of how CDD fees are calculated by the MSC shipowner at the Polish ports (with the consideration of different fees that depend on the shipowner's priorities).

Considering a relatively high level of complexity in generation and calculation of fees in international container turnover in maritime transport and a necessity of evaluating the fee amounts a priori, indices come as extremely useful tools to pursue a proper pricing policy (and calculation of costs incurred by stakeholders of processes based on forecasting).

6.6   INDICES IN MARITIME CONTAINER TURNOVER

Considering international container turnover, in the analysis of current freight rates and their forecasting, relevant indices are applied. They come as collective indices of freight prices/rates in the market of maritime container shipping and rates charged for container handling operations taking place at ports. Indices provide information on economic conditions and lines of economic changes in international container turnover. In the market of maritime container shipping, quite differently than in bulk cargo transport, there is not any single, universal freight index. Instead, numerous coefficients are applied to analyze the amount or dynamics of changes to charter and freight rates applied in the discussed market.

To analyze charter rates in container turnover, the following indices are used (Grzelakowski 2013, 128):

- ClarkSea
- Howe Robinson Container Index

- Harpex
- Maersk Broker Container Index
- Braemar BOXi Index
- Containership Time Charter Assessment Index

To analyze freight rates, the following indices are applied (Federal Reserve Bank of Dallas 2010):

- China Containerized Freight Index
- Shanghai Containerized Freight Index
- Drewry Container Freight Rate Index

The Clark Sea Index has been developed by M. Stopford and C. Tyler, and it has been published since 2002 by a renowned British brokerage company, Clarkson Research. The index is based on the average weekly charter rates/average profit calculated in USD thousand for one-day chartering of a relatively modern commercial vessel of any main type, including not only container vessels but also tankers, bulk carriers and gas tankers. The Clark Sea Index is weighed by the number of vessels operating in the particular segments of the charter market at the beginning of the calendar year. It is published once a week, based on the average charter rates calculated for the particular types of vessels in the previous week (The ClarkSea Index 2021). In 2019, the index was increased by 24%, reaching the level of 61.9 points at the end of the year. Then, in the first half of 2020, it was decreased as a result of a shock caused by the Covid-19 pandemic. The Clarkson company also publishes rates for chartering the particular types of container vessels, stated in USD thousand per day. From the beginning of 2020 until May 2020, these rates dropped down from USD 29,600 to USD 15,400 for vessels of the capacity of 8,500 TEU—that is, by 48%. For smaller vessels (up to 2,000 TEU), the rates decreased from USD 9,350 to USD 6,900—that is, by 24% (ISL Shipping Statistics and Market RevieI 2020, 9).

An index that is the most commonly applied one in container shipping is the Harper Index, published by a German brokerage company from Hamburg since 2004. At present, it is calculated based on daily charter rates, stated in USD, for nine types of container vessels, which are different in terms of their maximal container loading capacity (ranging from 700 to 8,500 TEU) or onboard cargo handling equipment. Harper Petersen & Co obtains charter rates indispensable to calculate the index from cooperating ship brokers, ship operators and owners who charter their vessels. Vessels that are chartered with the crew for 3 up to 48 months are considered in this calculation in terms of real demand and real supply of container vessels exclusively (Harpex 2021a).

The Harpex Index is based on the charter rates from 1986 at the level of 800 points. The changes to the values of the index in time depend on the demand for general cargo that dominates in container shipping. This demand, in turn, comes as a function of consumers' demand for the finished goods and companies' demand for semiproducts and equipment indispensable for production, which are all transported in containers by sea (Wagner 2014, 191). The more containers need to be transported by sea, the higher demand for container vessels and the higher rates for vessel chartering. The Harper Index is considered to be a reliable indicator of global economic conditions.

An increase in this index indicates an increase in global trade, whereas any decrease signaled by the index means falling tendencies in global trade volumes. The discussed index is also positively correlated with stock exchange indices and negatively correlated with bond indices. The values of the Harper Index undergo considerable fluctuations. At the beginning of 2002, the value of the index was 485 points, and it reached its highest level on 19th March 2005: 2,183 points. As a result of the global financial and economic crisis, at the end of 2009, the value of the index dropped down to 317 points, and on 22nd January 2021, it reached the level of 1,122.87 points (Harpex 2021b).

Published since 2011 by Drewry, a consulting company based in London, the World Container Index (WCI) is available every week. The index is based on container freight rates, including various additional fees (booking fees, customs fees, etc.) charged for shipping one 40' container along the eight most important shipping routes to/from USA, Europe and Asia: Shanghai–Rotterdam–Shanghai, Shanghai–Genoa–Shanghai–Los Angeles, Shanghai–New York, New York–Rotterdam–New York. The group of entities who systematically provide the Drewry company with the information about freight rates includes shipowners, customers using transport services, forwarders and other agents participating in maritime container turnover (Marine Link 2021).

As other indices, the WCI also undergoes constant fluctuations caused by such factors as macroeconomic and geopolitical uncertainty in the world, a gradual change in the size of container vessels in pursuit of economies of scale, a surplus in the supply of vessels, etc. Shippers' growing demand for container shipping services leads to an increase in the WCI, whereas a growing supply of transport capabilities of the global fleet of container vessels results in a decrease in the value of the discussed index. At the beginning of April 2020, the WCI was USD 1,526 per a 40' container, which is almost the same as at the beginning of the index on 23rd July 2011. At the end of January 2021, the index reached its highest level of USD 5,252/40' container (Drewry 2021).

The Freightos Baltic Index (FBX) has been published daily since 2018 by the Baltic Exchange in London, in cooperation with Freightos, a technological company based in Hong Kong. It has been quickly recognized as a reliable freight index in global shipping. The index is based on actual container freight rates provided on regular basis by shippers, forwarders and carriers to the Freightos company and its digital platform. The presented rates refer to 12 key container shipping routes that connect China/East Asia with both Americas, China/East Asia with Europe and Europe with both Americas. Each individual container shipping route is represented by five or seven main ports in each region. The FBX index presents an average of freight rates applied along the most important shipping routes (Freightos Baltic Index 2021).

Established in 2009, the Shanghai Containerized Freight Index (SCFI) is a Chinese index of container shipping rates. The SCFI is based on freight rates for shipping containers, stated in USD/TEU, that is applied along 15 shipping routes connecting Chinese ports with important foreign ports. The data are provided by key shippers and shipowners.

Another index in global container turnover is the New ConTex, published by the Hamburg and Bremen Shipbrokers' Association. It is based on daily rates charged for chartering container vessels of six types and the capacity ranging from 1,100 TEU to 1,700 TEU, for chartering container vessels of the capacity of 2,500 TEU and 2,700 TEU for one year and for chartering container vessels of the capacity of 3,500 TEU and 4,250 TEU for two years (New CoTex 2021). The level of the New ConTex index is presented in Figure 6.1.

The volumes of containers handled at maritime ports in the world are properly reflected by the RWI/ISL Container Throughput Index. The index is based on container handling operations performed at 91 ports, representing 60% of global container handling. The primary value of the index is based on the container handling operations performed in June 2015, assuming them as the level of 100 points. The development of the discussed index during the years 2019–2020 is presented in Table 6.7.

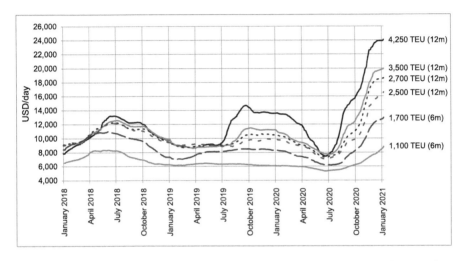

FIGURE 6.1 The levels of the New ConTex index presenting rates for chartering container vessels from January 2018 to January 2021 (USD/day).

Source: Data from Institute of Chartered Shipbrokers (2021).

TABLE 6.7
The RWI/ISL Index during the Years 2019–2020 (June 2015 = 100)

2019 (months)	RWI/ISL	2020 (months)	RWI/ISL
I	115.4	I	113.4
II	113.1	II	104.9
III	116.1	III	111.7
IV	117.2	IV	109.5
V	117.2	V	108.7
VI	117.4	VI	111.9
VII	117.0	VII	115.1
VIII	116.2	VIII	116.6
IX	115.5	IX	118.7
X	114.8	X	121.8
XI	113.7	XI	121.1
XII	112.9	XII	110.0

Source: Data from Statista (2021b).

As presented previously, the components of costs related to shipping and handling containers in international maritime trade clearly indicate the complex character of those processes. This fact, in turn, determines the necessity of applying uniform and transparent rules also in the field of documents required in international container shipping by sea.

NOTES

1. The INCOTERMS rules were developed in the 1930s as regulations defining terms of trade between the seller and the buyer. At present, they are most often applied to determine operations between the importer and the exporter, although they can be also applied in domestic deliveries. The rules were developed by the International Chamber of Commerce in Paris in order to maximally limit the number of disputes between suppliers and receivers. The INCOTERMS are regularly updated and the latest update took place before 2010. Their current version has been applied since 2020.
2. The Combiterms rules were first developed in 1949 by the International Chamber of Commerce in Paris (ICC) and the International Federation of Freight Forwarders Association (FIATA). They include commercial terms referring to transport of containerized goods and combined transport (forwarding collective shipments).
3. Elaborated based on E-logistyka (2021).
4. The interpretation of the *on board* term becomes even more complicated. In the European version of the Incoterms trade rules, it is explicitly translated as the board of a vessel (hence, the destination place is obviously a sea port/a container terminal), whereas in the America RAFTD 1941, the term may also refer to the railway car board, etc. (so a destination place is a railway station respectively, etc.). For example, a close interpretation of the European FOB rule in the RAFTD is the E-FOB rule, namely FOB Vessel.
5. More in Hermanowski (2004), Blajer (2000).
6. Tariffs charged for services provided by the WUŻ—Przedsiębiorstwo Usług Żeglugowych i Portowych Gdynia Sp. z o. o. (WUŻ Port and Maritime Services Gdynia, Ltd).
7. In Poland, TIR carnets are issued exclusively by the International Road Transport Association.
8. Volets (vouchers) are pairs of documents in one cover. One pair of documents is needed to cross one customs border. All of them must be filled in duly. In Poland mainly 6- and 14-volet carnets are available. At present, a 6-volet carnet costs PLN99.90 (2020).
9. If the goods take part in foreign fairs or exhibitions or if the samples of goods are to be presented to a foreign business partner, an ATA (Temporary Admission) carnet is applied to avoid customs procedures.
10. A SAD is a universal statistical customs document that is applied in trade with the countries from outside the EU. It refers to all goods. At present, it is applied in an electronic version and is sent to the customs systems as a document signed with the secure data transmission key.
11. A packing list includes data referring to each item on the invoice: the number of packaging units, the type of the packaging, the net and gross weight of each item and a summary.
12. A summary declaration is a document that includes all the information, based on which it is possible to identify the goods in transport. The customs services may check whether the particular goods can be imported to the country and whether they pose any threats. The standards referring to goods have been jointly developed for all the EU member countries. A summary declaration also allows the interested parties to make a quick decision whether to inspect the consignment or not. A summary declaration must be submitted at

the border customs office through which the goods are going to be transported. It must be submitted by a party who is going to transport the goods or a party who is responsible for the entire transaction. The scope of data included in the summary declaration depends on the mode of transport applied for shipping. Additionally, in the process, a temporary storage declaration is applied (during the customs clearance proceedings).

13. If the invoice does not come as a specification of goods, it is necessary to provide the information about the goods (a goods specification card), which may include a specification of goods or a list of goods, documents required for application of preferential tariff arrangements, a permit or any other document required for the export of goods, documents required to define the basis on which the taxable amount of goods can be determined if the invoice or any other document for the calculation of the customs value of the goods does not provide data indispensable to define such a basis; also, documents required to provide tariff classification of goods (such as certificates, issued by manufacturers or authorized scientific center, which provide the chemical and material composition of the goods up to 100%), information about the manufacturing process, an opinion of an expert, a specification defining the structure and functionality of a device.

14. The EORI (Economic Operators Registration and Identification) number is applied to identify business entities in operations involving customs and tax bodies in the entire EU territory. All importers and exporters who trade with countries from outside the EU are obligated to use EORI numbers. The EORI number is assigned only once, and it is unique for the entire EU territory. Since August 2015, it has not been possible to submit paper application forms for EORI numbers. Now it must be done only through the PUESC portal. At present, the EORI is legally based on the art. 9 of the Regulation no. 925/2013 of the European Parliament and of the Council, of 9th October 2013 on the Union Customs Code.

15. AES—the Automated Export System (in Poland, it is an IT tool for efficient prevention of pathologies that can be observed in the field of taking goods out of the EU customs territory), which refers to the electronic handling of export operations with the use of the e-Cło software, under the shared validation of XML documents; it is a fully digitalized service of certifying the EU status of goods (e-Status service), the beneficiary of which is the National Tax Administration.

16. The Provincial Inspectorate of Commercial Quality of Agricultural and Food Products.

17. The State Sanitary Inspectorate.

18. The State Inspectorate for Plant and Seed Protection.

19. Customs duties and taxes must be paid within ten days, starting from the acceptance of the customs declaration by the customs office; the payment is made by the importer while the documents are most often prepared by the customs agency. The customs office admits the imported goods for trade only after the customs duties and taxes—calculated based on the submitted customs declaration—are duly paid.

20. The word *freight* has more than one meaning. It may refer to the transportation of goods not only by sea but also by railway or by passenger cars. It may also refer to cargo or a freight bill.

21. Bunker is fuel oil used for propelling the main engines of a vessel.

22. It results from the obligation of transferring empty containers from the place where a surplus of containers can be observed to the place where there is a shortage of containers (repositioning).

23. This fee is practically charged by container carriers with the freight and then settled with the terminal operator.

24. Ibidem, pp. 267–268.

7 Documents in International Maritime Transport

7.1 A BOOKING CONTRACT AND A BOOKING LIST

Most often, maritime container shipping takes place regularly and is implemented with the use of container shipping lines. A shipping line is usually a fleet of vessels that provides shipping connections between particular seaports (actually—between particular container port terminals) (Kujawa, ed. 2015, 245). Shipping takes place in accordance with sailing lists, namely, timetables of calls of vessels at the particular ports located along a particular shipping route.

A maritime shipping line carrier is obligated to provide their services during the entire time period of the line operation and to inform about any changes to the sailing lists or any suspended sailings duly in advance. The operation of shipping lines comes as shipping companies' commercial offers, which can be provided to any interested parties, based on the public access (common carrier, public carrier) principle. Hence, cargo consigners can choose a shipping line that is most convenient to their needs. Considering directions of shipping and geographical regions, it is also possible to use various occasional shipping bargains that sometimes appear on the market.

In line shipping, there are various shipping contracts applied; however, the most common one is a booking contract (booking notice, booking confirmation). In accordance with the Maritime Code, a booking contract refers to transportation of particular cargo described in the contract in terms of its type and external dimensions (*Journal of Laws* 2018b). The core of a booking contract is to achieve its purpose, namely to deliver the cargo; the contract itself is not strictly related to a particular vessel, as opposed to charter contracts (Łopuski, ed. 1998, 327). The fundamental responsibility of a maritime carrier is to transport the cargo contracted by the booking party (Kunert 1970, 287–288). The booking party is obligated to pay the freight fees for transportation of the cargo and to make sure that the cargo is delivered at a particular time to a particular seaport, considering a particular sailing list (at the particular container port terminal). Usually, based on the booking contract, containers and general cargo unitized in various forms are transported; however, it is also possible to transport bulk or semi-bulk cargo.[1]

The public character of a booking contract results in the fact that a particular line vessel is loaded with goods from various booking parties (Kujawa, ed. 2015, 248). The line carrier should provide proper shipping conditions to all their customers: to provide space for their cargo on the vessel and to deliver the goods in accordance with the current sailing list. Carriers who provide transportation services are interested in using their loading space and/or loading capacity in an optimal way and also in providing vessels that are capable of delivering proper service (the carrier can change line vessels according to their needs).[2]

Information about the availability, operating regulations of a particular shipping service and terms of transportation are respectively provided in a sailing list, a current freight tariff, a booking contract (if in writing) and a liner bill of lading. The shipping route is determined by the sailing list, which provides all the required information indispensable to plan cargo transportation. The sailing list provides the following information:

- the type of a shipping line
- all the ports located along the shipping route where the vessel calls at or, if there is such an option, all the additional (range) ports that can be called at under certain conditions
- the frequency of calls made by the line vessels at the particular ports along the shipping route
- information about joint line services if the carrier cooperates with other shipowners
- information about feeder services to the particular ports along the shipping route

The key information for the booking party is the type of the shipping line (a container line that can handle rolling cargo, a ferry line that can transport passengers and cargo) because it indicates the type of goods that can be accepted for transportation. It is important in terms of adjusting the vessel to a particular type of cargo and adequate cargo handling equipment at the port terminal. The consigner or the forwarder of the cargo knows the specific character of their cargo, and they have to select an adequate shipping line (they must stuff containers, load them on semitrailers, low-chassis semitrailers, Big Bag sacks or any other loading units accepted by the carrier for transportation).

Based on the previous information, the booking party can plan the shipping route for their cargo considering a seaport located in the nearest vicinity of the cargo destination place. If the sailing list contains an option of calling at an additional (range) port, the booking party should agree on terms for such a sailing with the carrier.[3] Furthermore, a sailing list usually provides the particular dates or even hours of calls. This allows the interested parties to define an approximate time of the entire transportation process between the particular ports. All the information about the current position of the vessel, previous calls made at other ports and the identification data of the vessel can be obtained through the navigation monitoring applications (e.g., vesselfinder.com; marine traffic24.com).

Joint shipping services are provided by several maritime container carriers in order to increase their service range (Kujawa, ed. 2015, 104). The higher number of vessels providing services along a particular shipping line allows carriers to increase their service supply and to attract new booking parties. The cooperation allows them to improve the availability of their services and reliability of container shipping (to guarantee that the service will be duly provided).

Carriers state their provisions referring to the booking contract on its form (if it is in writing) or/and in a liner bill of lading. The booking contract and the liner bill of lading state, among others, the following: terms of transportation, the range and basis of the carrier's liability for the transported cargo, freight payment terms, lien on the cargo and proceedings applicable in various situations referring to sea journeys.

Entering into a booking contract may take place in accordance with local customs/traditions or in a form of a verbal submission of the cargo for transportation (e.g., via a phone call). It can also take the form of a written submission, it can be sent by fax with all the required information provided or it can take place by filling in a form on the carrier's website or a website of the carrier's agent.[4] The booking party may receive a confirmation in the form of a position number defining the cargo on the booking list. The booking contract may contain only the information necessary to identify the consignment and to provide the shipping service, namely the following (Kujawa, ed. 2015, 347):

- identification of the booking party
- identification of the line carrier
- information about the type and dimensions of the cargo
- information about the shipping route (the port of loading the cargo and the port of destination)
- information about the freight

For the booking party, it is important to know the limitations referring to the transportation of the particular cargo groups—it mainly refers to dangerous cargo and goods that require special transportation conditions (e.g., special monitoring, cooling, oversized pieces that exceed the dimensions of a typical ISO container). There are also situations in which the booking party submits a large number of loading units for transportation but the whole lot cannot be transported during one sailing because the carrier has to consider other shippers' cargo as well (and to consider the order of cargo submissions). It is also important to state clearly and precisely the type and the name of the cargo as it is required for any shipping operations involving dangerous goods that undergo the respective classification, depending on their physical and chemical parameters and the level of hazard they pose to the environment.

The requirements referring to the classification, marking, packaging and transportation procedures are precisely regulated by the International Maritime Dangerous Goods Code—IMDG) (International Maritime Organization 2021). The booking contract defines legal relations between the booking party and the maritime container carrier, and its terms are stated in the liner bill of lading (which is also a shipping document, although not a shipping contract). In economic practice, it is possible to find standard forms for booking contracts, such as *Conline booking*, or general forms for booking contracts, such as *Blank back form of liner booking note*.[5]

As a result of entering into a booking contract, a necessity arises to provide a booking/loading list. It is usually prepared by the booking office of the shipowner or the shipowner's agent. A booking list is a list of goods contracted for transportation during a particular sailing, by a particular vessel. The list must be provided to all the interested parties before the vessels arrives at the port for loading. Based on the loading list, a stowage plan is developed, stating precisely how the cargo must be loaded and distributed on the vessel. With a booking list, it is always possible to check whether all the goods have been duly loaded. It can be used for inspections of the cargo that is transported by the vessel. The booking list is also a document to confirm the acceptance of the particular goods on the board of the vessel, and in the case when there is no written confirmation of the entering into a booking contract, it comes as an evidence that there has been such a contract concluded. In fact, the entering of the cargo submitted by a particular shipper into the booking list comes as a confirmation of the entering into a contract. The shipper or their representatives should be notified about that fact.

7.2   A SEA WAYBILL

Typical shipping documents applicable in maritime transport of general cargo are a booking list and a liner bill of lading. However, these documents entail a necessity of meeting some particular procedural requirements related to the possibility for the consignee to decide about the cargo. Releasing the cargo to the consignee is possible only after the consignee hands over the original bill of lading to the maritime carrier. As mentioned before, a bill of lading is a negotiable security that establishes the title to goods and can be transferred to a third party who becomes the owner of the goods specified in the document. At present, the possibility to transfer/sell a bill of lading to a third party is very rarely used, and shippers are more interested in providing efficient deliveries to destination ports. In this situation, the fundamental feature of a bill of lading is not used, and the document itself is applied as a shipping document. Hence, an idea has appeared to implement a simplified shipping document, without the feature of a negotiable security establishing the title to the goods, next to the original bill of lading.

A sea waybill (a seaway bill, an express cargo bill) is applied as a confirmation of a shipping contract that has been entered with a shipping company, mainly to transport maritime ISO containers. It is also proof that the particular goods have been contracted for transportation and received by the carrier (Łopuski, ed. 1998, 413–422). If a sea waybill is applied in shipping, it actually stands for two documents, namely a booking list and a liner bill of lading. Based on a sea waybill, the carrier is obligated to transport the cargo and to release it to a person stated by the consigner as a consignee of the cargo. The document itself does not bear any features of a negotiable security establishing the title to the goods, and there is no possibility to sell it to any third parties. A sea waybill is a document issued specifically to the name of the consignee of the cargo; therefore, there is very little risk that the cargo is going to be released to an unauthorized party. The delivery of the cargo is a quick process that usually takes place without any problems. The receiver of the goods does not have to have the original document to receive the cargo (as opposed to the case involving a bill of lading, where all the original documents must be presented).

The shipping line carrier issues a sea waybill at the port of loading and hands it over to the shipper. The number of the document copies is not limited, and there are no specific regulations to define it precisely. It is possible to indicate the consignee of the cargo at the end of the shipping process, at the destination port. It is even possible to indicate a consignee different from the initially defined one (Kujawa, ed. 2015, 301).

The structure of A sea waybill is based on a bill of lading—that is, it states terms of transporting the cargo, including the scope of the liability for the cargo, the way of shipping and methods of calculating the freight fees. In the section where the identification data should be provided, it is necessary to specify the consigner, the carrier, the consignee of the cargo and the characteristics of the cargo. The information stated in a sea waybill is used by carriers to generate numerous auxiliary and additional documents, such as cargo and customs manifests, booking lists, vessel arrival notifications, freight bills, manipulation orders for port terminals (Kujawa, ed. 2015, 300). The flow of documents is frequently supported by modern IT systems that provide adequately modified data indispensable for issuing relevant documents, including sea waybills.

As opposed to a bill of lading, a sea waybill does not provide as many possibilities to the consigner and the consignee of the cargo. The consigner cannot use the sea waybill as a confirmation of meeting the delivery terms to the consignee's bank (the buyer who pays for the transaction) because this kind of document is not accepted for bank settlements. The consignee of the cargo may decide about the cargo at the moment when they receive it from the carrier, but during the transportation, the consignee does not have any rights to the cargo. The lack of a possibility to sell a sea waybill is, on one hand, its biggest limitation, but on the other hand, it streamlines the flow of documents and quick reception of the cargo at its destination port. Despite its limitations, a sea waybill is more and more commonly applied in maritime container shipping.

In 1990, some regulations referring to sea waybills were implemented: CMI Uniform Rules for Sea Waybills 1990, developed by the International Maritime Committee (CMI), UNCTAD and the Economic Commission for Europe. The norms clearly indicate that a sea waybill does not have any attributes similar to a bill of lading, and therefore, it is not applicable under the Hague-Visby Rules. However, the legislation of the particular country may determine the liability of carriers based on its national standards. Frequently, domestic standards include references to the international regulations that are the basis for the carrier's liability for any damage to the cargo. Considering a bill of lading, the Hague-Visby Rules are most often applied. The liability of the carriers who issue sea waybills and bills of lading have often been the subject of judicial proceedings, and as a result, the application of the Hague-Visby Rules has been recommended, such judicial decisions have been issued in Great Britain, the United States of America and Canada (Pamel 2011, 16).

Considering its attributes, a sea waybill is commonly applied by international companies that run their manufacturing subsidiaries in the industrial production centers based in the Far East. Sea waybills are applied by importers who import large quantities of various goods from Asia to Europe and to the USA. Considering document handling costs (fees charged for issuing documents), a sea waybill is more and more applied, especially in the shipping of low-value goods, such as textiles, small consumer goods, some household appliances, inexpensive electronic goods and toys.

7.3   A LINER BILL OF LADING

Similarly to a situation when non-containerized cargo is transported, to manage and to control the process of shipping containerized cargo properly, relevant documents are required. Their proper development is of key significance to the course of the entire shipping process. The most important document applied in maritime trade is a bill of lading (B/L). In regular line shipping, a bill of lading (a liner bill of lading, a liner B/L) is issued after the booking contract has been entered by the owner of the cargo, its consigner or a forwarder, acting on their behalf as a consigner, and the shipowner or the shipowner's agent.

A booking contract regulates legal relations between the consigner and the shipowner whereas a bill of lading defines legal relations between the shipowner and the consignee of the cargo. A bill of lading indicates a maritime carrier and a consignee authorized to decide about the cargo. The carrier is obligated to deliver the cargo to its destination port against an agreed payment and to release the cargo to the authorized consignee in exchange for all the original documents (Figure 7.1).

A bill of lading is filled in by a forwarder and signed (issued) by the shipowner (the captain of the ship or an authorized officer). Depending on whether the bill of lading proves the actual loading of the cargo on a particular vessel defined by name or whether it only confirms the fact that the cargo has been received for transportation, it is possible to determine two types of bills of lading (Łopuski, ed. 1998, 391):

- A shipped-on-board bill of lading—annotations referring to any possible damage are filled into the bill of lading with the indication of the particular lot or pieces of the cargo. After such annotations are made, the document becomes a dirty bill of lading. A dirty bill of lading with such annotations might not be accepted by the bank for financial settlements, and the price of the cargo delivered to the consignee becomes lower in the case of sale.
- A received-for-shipment bill of lading—after the cargo has been loaded on the vessel, the document is obligatorily exchanged for a shipped-on-board bill of lading. There is no date specified in the document to indicate the delivery date to the destination port of the cargo. The commercial value of this type of document is lower than of a shipped-on-board bill of lading because in most documentary letters of credit, a shipped-on-board bill of lading is required to confirm the actual loading of the goods on a particular vessel.

FIGURE 7.1   Legal relations between the participants of maritime container shipping.

Source: The authors' own elaboration.

A bill of lading is issued by the shipowner upon the shipper's demand (the consigner of the cargo, who performs or supervises the process of vessel loading). The issuance of a bill of lading is not obligatory, and it is possible that such a document is not issued at all. However, if a bill of lading is issued, it is issued in sets. A set usually contains three originals and several copies, which do not entitle their owner with any rights. The forms of a bill of lading applied in international trade by various shipowners are not generally standardized; however, their content is very similar. The only difference is the arrangement of the information included in those documents. A bill of lading should include the following data (*Journal of Laws* 2018a):

- information about whether the document is a shipped-on-board bill of lading or a received-for-shipment bill of lading
- the name and the address of the consigner of the cargo, who is usually its exporter
- the name and the address of the carrier
- identification of the consignee or an annotation that the bill of lading is issued to an order or to the bearer
- the name of the vessel
- information about the type of the cargo
- information about the condition of the cargo
- the cargo identification markings
- specification of the freight fees and any other receivable payments for the carrier or an annotation informing that the freight fees have already been duly paid
- the port/place of loading and unloading the goods
- information about whether the bill of lading is the original copy and how many original copies of the document are provided
- the date and the place of the issuance of the bill of lading
- the signature of the carrier—the captain of the vessel or an authorized officer

A bill of lading is a negotiable security, representing the cargo described in the document and the cargo can be released to its consignee only against this particular bill of lading. The consignee of the cargo must have the complete set of documents and to meet the terms of payment duly. Usually, a bill of lading follows the procedures listed as follows:

1. After the goods are loaded onto the vessel or after the goods are received for shipment, the bill of lading is handed over to the cargo consigner (often by a maritime forwarder).
2. The consigner of the cargo sends the bill of lading by courier mail to the consignee of the cargo (the bill of lading is never sent with the cargo by sea).
3. At the destination port, the consignee of the cargo (usually with the assistance provided by maritime forwarders) presents the complete bill of lading set.
4. The carrier releases the cargo to its consignee.

Apart from the previously mentioned bills of lading, their following types can be also observed in maritime shipping: direct (through), multimodal and FIATA bills of lading.

7.4 DIRECT (THROUGH), MULTIMODAL AND FIATA BILLS OF LADING

Apart from liner bills of lading, in containerized cargo shipping, there are direct or through bills of lading applied, which allow several carriers to transport the same consignment. A direct bill of lading or a through bill of lading covers several sections of a shipping route covered by cooperating carriers. The cargo may be transported by various means, often along the sea and land sections of the entire shipping route.

Direct/Through bills of lading can be divided into the following (Łopuski, ed. 1998, 471; Kunert 1970, 412):

- maritime documents that are issued and applied in maritime transport, where only seagoing vessels are used and there is a change of the carrier
- combined, where maritime transport covers only a particular section of the route and other sections are covered by land, inland–waterway and air modes of transport

Maritime through bills of lading are issued most often when there is no direct line connection from the initial port to the cargo destination port. In such a case, it is necessary to transport the cargo by feeder vessels to the nearest port that provides the required connection, where the cargo is unloaded at the port terminal and then loaded again on another vessel calling at the cargo destination port. Referred to as feeder service, such a solution is applied at smaller local ports.

If a feeder carrier and an oceangoing carrier both participate in the process of cargo shipping, each of them is entitled to issue a bill of lading; however, most often, the document is issued by the oceangoing shipowner. The issuer of the bill of lading is responsible for the due delivery of the transportation service along the entire shipping route (including cargo handling operations) until the moment when the cargo is released to its consignee (*Journal of Laws* 2018b). The issuer of the bill of lading may act as a contracting carrier—that is, to accept responsibility for the entire shipping route until the moment when the cargo is released to its consignee. In a situation where the carrier who issues the bill of lading is the actual carrier, the carrier transports the cargo themselves along one of the sections of the entire shipping route, and other sections are covered by other maritime carriers. When a through bill of lading is issued by a contracting carrier, then the responsibility is jointly accepted with other subcontractors (if the carrier does not provide the shipping service themselves) or cooperating carriers (if the issuer transports the cargo along a particular section of the shipping route). Another solution is to limit the liability of the issuer of the bill of lading only to the section of the shipping route covered by the issuer and to do it similarly for other carriers. Most frequently, each carrier is responsible only for the section of the shipping route covered by themselves, and the issuer of the bill of lading should secure any possible claims of the booking party if there is any damage to the cargo caused (art. 139 §2) (Łopuski, ed. 1998, 471–472). Claims for damages to the cargo mean that the consignee is required to prove where the damage was caused (along which section of the shipping route) and which section carrier is to be held liable for the damage. It comes as a serious inconvenience for the booking party and to the consignee of the cargo because in such a situation, they need to incur any costs related to the investigation and identification of the carrier responsible for the damage. Hence, the most convenient solution for the booking party (and for the consignee of the cargo) is when the issuer of the bill of lading assumes the liability for the cargo along the entire shipping route.

In feeder shipping, an additional type of through bills of lading is applied, usually referred to as a local bill of lading. This document does not have to be issued to the booking party (the consigner) but acts as a shipping document that regulates the transportation and reception of the cargo by the subsequent carriers. The local bill of lading defines the rules for shipping processes, whereas the cargo is released to the consignee against the direct/through bill of lading issued by one of the carriers.

Maritime carriers may also issue a joint bill of lading, which is one document for all the contractors of the shipping process. Based on a through bill of lading, all the carriers assume joint and several liability for the shipping service, and the clauses of the bill of lading precisely define relations between the particular carriers and the consignee of the cargo. As in the case of any other direct bills of lading, the carriers may be held liable only for the sections of the shipping route they cover (a disclaimer of joint liability, art. 138 § 2k.m.).

As mentioned before, mixed through bills of lading are applied in shipping whenever it is necessary to combine various modes of transport. This operation allows the interested parties to arrange a land–sea transport system. The development of containerization has resulted in a considerable interest in solutions that allow parties to organize shipping from the consigner to the consignee of the cargo (door-to-door). The liability of the particular carriers for the sections of a shipping route they handle is defined by specific sectoral regulations that are currently binding for the particular mode of transport. Considering maritime transport, these are regulations referring to the liability resulting from the issuance of bills of lading, namely the Hague-Visby rules of 1928 (with two amendments of 1968 and 1979) and the Hamburg rules of 1978. Considering other modes of transport, the current regulations can be found in the Convention on the Contract for the International Carriage of Goods by Road of 1956, the Convention on International Carriage by Rail and the conventions regulating the rules of air transport—the Warsaw Convention of 1929 (with two amendments of 1955 and 1971) and the Montreal Convention of 1999.

Considering mixed through bills of lading, the particular carriers of various modes of transport implement shipping processes along their sections of shipping routes, and they issue bills of lading in which the consigner of the cargo is the carrier who is also the issuer of the particular bill of lading. All carriers are responsible for the shipping process based on the issued shipping documents and in accordance with the current legal regulations specified for the particular mode of transport. If there is any damage to the cargo, the issuer of the bill of lading must secure the evidence for the claims filed by the consigner or the consignee of the cargo. The carrier who is held liable for the damage is the carrier on whose section of the shipping route the damage has been caused. The carrier is held liable in accordance with the regulations accepted for the particular mode of transport (transport network liability) (Szczepaniak, ed. 1996, 169).

To strengthen the legal position of consigners and consignees of goods, a new convention was implemented in 1980 to regulate multimodal shipping. The United Nations Conference on a Convention on International Multimodal Transport implements a multimodal document based on which shipping of goods can be provided. The convention states that a multimodal transport operator (MTO) is engaged and responsible for transportation along the entire shipping route, based on the single accountability rule—the operator is held liable regardless of the place where a damage has been caused to the cargo (UN Conference 1980, 5). The convention increases the scope of contracting carriers' liability (MTO), and it implements the principle of the carrier's fault presumption. This principle, however, has been strongly opposed by carriers, and some countries have refrained from the ratification of the convention. At present, there is not any comprehensive system of legal standards defining the principles for multimodal transport because the convention of 1980 has not been entered into force. However, there are other regulations provided by the following:

- The International Chamber of Commerce, regulations of 1973 (revised in 1975) and of 1992
- The United Nations Conference on Trade and Development and the International Chamber of Commerce, regulations of 1991.[6]
- The Baltic and International Maritime Council—BIMCO, regulations on the principles for issuing the COMBICONBILL, implemented in 1971; COMBIDOC, 1977; and MULTIDOC 95, 1995 (Łopuski, ed. 1998, 489; Młynarczyk 1997, 166; Kujawa, ed. 2015, 281).
- The International Federation of Freight Forwarders Associations (FIATA), regulations related to the issuance of a multimodal bill of lading FIATA FBL (1975); the latest amendment to the FBL was in 1992 (International Federation of Freight Forwarders Associations 2010).

The issuer's liability results from the clauses of a bill of lading, but it is also regulated by the local legal norms—e.g., the statute of limitation period for claims filed by the consignee is regulated by the rules of the Polish Civil Code and Maritime Code (Janicka 2012, 6). A multimodal bill of lading might be issued by a shipowner, a land carrier or a forwarder; however, the characteristic feature of this document is the fact that the liability for the entire shipping process—from the consigner to the consignee of the cargo—is assumed by the issuer. Hence, the issuer becomes a contracting carrier who can provide shipping services, cooperating with subcontractors who are most often carriers operating in various modes of transport. Multimodal transport can cover the entire shipping route without any maritime section at all, where only means of land or/and inland waterway transport are required.

A FIATA Multimodal Transport Bill of Lading, a negotiable FIATA Multimodal Transport Bill of Lading (FBL), is a document that takes a special place in transport.

It performs all the functions of a bill of lading issued by a shipowner—that is, it is a negotiable security. An FBL may be used by carriers associated in national organizations who belong to the International Federation of Freight Forwarders Associations (FIATA), and this principle is strictly followed (Janicka 2012, 4). The issuer of an FBL is required to hold indemnity insurance against liability arising from the risk related to the organized shipping process. As in any other shipping processes performed with the use of multimodal bills of lading, the issuer is held liable in two ways:

- Based on the transportation network principle—if it is possible to determine the place where the damage to the cargo has been caused, the operator (MTO) is liable in accordance with the principles applied in the particular mode of transport respectively.
- Based on uniform terms—if it is not possible to determine where the damage to the cargo has been caused.

The entity issuing an FBL as the carrier assumes the liability for the entire shipping route of the cargo, based on the principles typical for a contracting carrier. If there is any damage caused to the cargo, the carrier is held liable depending on the place where the damage has been caused. If the cargo has been lost or damaged during the transportation, the MTO or their insurer has to pay compensation; however, if they have cooperated with subcontractors, they will file recourse claims to the respective carrier (Janicka 2012, 7). The FBL document is also applied in trans-Atlantic shipping, and it follows the principles of liability referring to maritime transport to and from the United States of America, namely the Carriage of Goods by Sea Act (COGSA) (Sweeney 1999, 580–581). The carrier often acts as a non-vessel operating common carrier (NVOCC), a transport operator without their own vessels who offers shipping services on their own behalf. An NVOCC operator enters into a contract for the provision of space for the transportation of containers with a sea carrier whose vessels operate along particular shipping routes. Entering into contracts with several shipowners allows an NVOCC operator to offer shipping services on the most frequently used shipping routes.

The implementation of the FIATA document with the complete attributes of a bill of lading has allowed land forwarders to issue documents equivalent to classical maritime bills of lading and to become involved into organization of shipping processes of the global range. Carriers (MTOs-NVOCCs)[7] are now able to provide shipping services to their customers in relations where various ways of transporting the cargo are applied with the use of various modes of transport.

7.5   A SLOT-HIRE CONTRACT

Multimodal transport operators (MTOs) and those who do not have their own vessels (NVOCCs) are interested in hiring loading space for their cargo on vessels belonging to actual carriers, namely those who operate their own sea-going vessels. An agreement that is most frequently applied to hire loading space on a vessel is a slot-hire charter contract (Kujawa, ed. 2015, 225). Based on such a contract, the chartering party (MTO, NVOCC) gets an opportunity to transport their own cargo or their customers' cargo on the provided vessels.

The essence of a slot-hire charter is hiring individual container slots—where a slot is a space where a 20′ container can be loaded. Shipowners let their loading space for a defined period of time and a defined number of container slots (TEU), and additionally, they may specify the maximal weight of containers and the shipping routes for their vessels (Łopuski, ed. 1998, 595). Depending on the chartering party's requirements, the slots can be equipped with power connections for shipping reefer containers. In the contract, the shipowner and the chartering party agree upon a possibility of transporting dangerous or atypical cargo in containers. The containers transported on the vessel must meet the ISO standards and all the requirements related to transportation safety (the Convention of 1972).

MTO and NVOCC operators provide their customers with services of shipping cargo on their own behalf, and they issue relevant documents that are most often sea waybills, liner bills of lading and multimodal bills of lading. Consigners enter into shipping contracts with operators, and there is no need to contact actual carriers. MTO and NVOCC operators run all their financial settlements with their customers, including customer complaints.

Under the terms of a slot-hire charter contract, the chartering party may hire container slots for a single sailing or several sailings or agree with the shipowner upon a specific time period of the contract. A slot-hire charter contract usually specifies a vessel or vessels on which container slots are chartered, the precise number of slots, the shipping route, the shipowner's remuneration and other significant information, including regulations for the parties' liability for any damage or loss of the cargo. The freight fee is paid by the chartering party, and it depends on the number of chartered slots. It is usually paid for a TEU. The shipowner calculates the price for a TEU depending on their fixed and variable costs, including additional costs that are incurred when, for example, reefer containers are transported. Most often, MTO and NVOCC operators enter into slot-hire charter contracts with shipowners who operate on container shipping lines and provide a network of regular shipping connections. In this way, they can implement transportation processes between major seaports in the world.

Before transportation of the chartering party's containers, the actual carrier issues a master bill of lading that includes all the containers contracted for transportation by the MTO or NVOCC. The document comes as the basis for financial settlements between the chartering party and the shipowner, and it is referred to whenever it is necessary to establish liability for any damage of the goods. The contracting carrier—that is, an MTO or an NVOCC—is liable for damage of the cargo to the shipper and the consignee of the cargo. Based on the slot-hire contract and the master bill of lading, the contracting carrier may file recourse claims to the shipowner, who is the actual carrier.

The dynamic development of container transport is a factor that has accelerated the implementation of flexible organizational solutions and the provision of adequate shipping documents. In 1993, the Baltic and International Maritime Council (BIMCO) implemented a specimen of a charter contract—a standard slot charter.

7.6   THE SIGNIFICANCE OF A BILL OF LADING IN INTERNATIONAL SETTLEMENTS WITH THE USE OF A DOCUMENTARY LETTER OF CREDIT

A bill of lading plays an important role in payment for goods imported by sea with the use of a documentary letter of credit. In this situation, the fact that a bill of lading is a document confirming the right for the goods upon which it has been issued is referred to. The consignee of the bill of lading (who has all its copies) issued by the carrier is entitled to decide about the cargo during the transportation until the moment when the cargo is received after the vessel has arrived at the destination port.

In the pragmatics of processes, "a documentary letter of credit comes as an obligation of the bank to the exporter, undertaken by the bank upon the importer's request, to pay and to secure payment for the exporter, at a specified amount, in return for submitting the documents specified in the letter of credit that meet all the requirements defined in that letter of credit, within the specified period of time" (Marciniak-Neider 2008, 77). A documentary letter of credit is opened for the benefit of the exporter by the importer's bank, upon the request of the importer who submits or secures the financial coverage of the documentary letter of credit. Then the importer's bank informs the exporter's bank, which acts as an intermediary bank, about the opening of the documentary letter of credit, and the exporter's bank informs the exporter about this fact. Knowing that the documentary letter of credit has been opened for them, the exporter transports the goods, completes the required documents in accordance with the terms specified by the documentary letter of credit and submits them in the exporter's bank. The exporter's bank sends the documents to the importer's bank. Having verified the conformity of the documents with the requirements stated in the documentary letter of credit, the importer's bank transfers the payment to the exporter's bank. The importer's bank sends the documents to the importer to allow them to receive the goods, and finally, the exporter's bank transfers the payment to the exporter.

Considering maritime transport, one of the documents indispensable for payment in the form of a documentary letter of credit is a bill of lading, whose content must be consistent with the terms of the documentary letter of credit. Additionally, a bill of lading may be applied as an instrument to secure a credit granted to the importer by the importer's bank. Such a situation may occur when the importer's bank opens a letter of credit without its financial coverage on the importer's bank account. It means that the bank that opens the letter of credit grants a credit to the importer and then might have some difficulties to recover the funds paid to the exporter for the delivered goods. In such a situation, the structure of a bill of lading allows the importer's bank, which has got the bill of lading related to the particular documentary letter of credit, to make it impossible for the unreliable importer to receive the goods (Szumański, ed. 2007, 294).

7.7   A CARGO MANIFEST

In maritime shipping, including container shipping as well, each captain of a commercial vessel must be provided with an important document, namely a cargo manifest (a ship cargo manifest). The document can have two forms: it can be a loaded cargo manifest or an unloaded cargo manifest. A cargo manifest is issued at the port where the cargo is loaded on a vessel, and then it becomes a loaded cargo manifest. At the destination port, the same document becomes an unloaded cargo manifest.

A cargo manifest is a list of the particular lots of goods loaded onto the vessel by the order of calling at their destination ports. At each loading port, a separate loaded cargo manifest is prepared for the cargo that is to be transported to the particular destination port. The manifest is based on special forms and on the data provided in the bills of lading or sea waybills. The document is signed by the carrier's representative or agent. The copy of the loaded cargo manifest is submitted to the customs bodies before the departure of the vessel.

A cargo manifest is a register of consignments loaded onto a vessel for transportation. Each consignment registered in the cargo manifest has got the number of the bill of lading and is described with the following data: the name of the consigner, the name of the consignee, the identification of the goods, the number of pieces, the type of the packaging, the gross weight, the capacity of the goods, the freight rate, the freight amount paid and due and other annotations on the condition of the goods. Knowing the number of the bill of lading and having access to the IT system of the shipowner, it is possible to monitor the movement of the particular full containers from the loading port to the destination port of the entire consignment. Then it is also possible to monitor empty containers coming back to the shipowner's depot. Based on the loaded cargo manifest, the shipowner's agent prepares an unloaded cargo manifest at the destination port of the cargo. This document is not different from the loaded cargo manifest unless, during the sailing, some identification data of the cargo have been changed—for example, when as a result of a storm, some containers have been washed off the board or have been unloaded at a different port by mistake (Marciniak-Neider and Neider, eds. 2014, 257).

Apart from the purpose of identifying the cargo that is to be unloaded at a particular port, the cargo manifest is used during the customs clearance procedures for import of goods by sea as a summary customs declaration. The cargo manifest is also applied for statistical purposes. It should be emphasized that in the era of digital economy, it is possible to develop a digital container identification card (referred to as a Unit Inspector tab) based on the cargo manifest. The card contains basic data about the containerized consignment, data informing about the condition of the container, the seal numbers and information about the customs status of the goods transported inside the container.

As previously discussed in this chapter, the problems referring to documents applied in international maritime container shipping have been focused on documents related to provision of transportation services. However, the entire process cannot be fully discussed without considering the procedures related to the insurance of containers and vessels in maritime transport.

NOTES

1. Most often, bulk and semi-bulk cargo are transported in Big Bags, which allow operators to handle the cargo with the use of mechanical devices. See more on Opakowania.com.pl (2021).

2. Unless there is a provision that forbids to substitute the vessel in the booking contract. In such a case, the carrier is not allowed to substitute the vessel with another one (and there is no such practice applied). Usually, there is an information note stating that there is a possibility to substitute the vessel with a similar one that provides the same transportation capabilities.

3. Additional (range) ports are considered in sailing lists, and they are called at under certain conditions—for example, where there is enough cargo to justify a longer sailing in order to cover an additional distance in economic terms. At present, distant ports are usually handled by smaller feeder vessels that connect them with big sea ports.

4. The booking contract term may refer to the entering into a verbal or written agreement whereas a booking note refers to a written confirmation of the contract. If there is no confirmation of the entering into a booking contract (e.g., a verbal declaration of will), it is assumed that the contract has been entered in accordance with the terms commonly applied by the carrier who has been selected to deliver the service.

5. These documents have been developed by the Baltic and International Maritime Council (BIMCO), and they are recommended for entering into booking contracts in their written forms.

6. The regulations referring to documents applied in multimodal transport implemented by the UNCTAD and ICC. See more in: (UNCTAD/SDTE/TLB/227 2001) *United Nations Conference on Trade and Development Implementation of Multimodal Transport Rules*, Report prepared by the UNCTAD Secretariat, UNCTAD/SDTE/TLB/227, June 2001.

7. Non Vessel Operating Common Carrier (USA), or less often, Non Vessel Owning Common Carrier.

8 Insurance of Containers and Vessels in Maritime Transport

8.1 CIRCUMSTANCES FOR INSURING CONTAINERS, CARGO AND CONTAINER VESSELS

Even if safety standards are strictly followed in maritime shipping, it is impossible to prevent all accidents and their results that can occur at ports and along shipping routes. Accidents can be caused by forces of nature (natural causes), or they come as the consequences of human actions (anthropogenic causes). At ports, vessels and cargo are threatened by such force majeure events as earthquakes, hurricanes, tsunami and floods. Threats resulting from human actions include the following (Tubielewicz and Forkiewicz 2013, 575–576):

- damage to the goods and various accidents during cargo handling and stowage operations
- terrorist attacks on port facilities that can be also destructive for the port town and its region
- accidents during operations of refueling vessels with bunker oil
- ecological threats related to cargo handling or storage of dangerous goods (oil spills, cargo, gas or ammunition explosions, fires involving tankers with liquid fuel, etc.)
- theft of goods at the port or destruction of the cargo as a result of an explosion
- threats related to drug and weapon smuggling that can disturb efficient cargo handling operations
- political events that can disrupt port operations, such as strikes, protests, demonstrations, port entrance or road blockades
- failures of IT and telecommunication port systems, hacker attacks

At sea, vessels are threatened by the following: war operations, pirate attacks involving the necessity of paying a ransom or, in the worst cases, a seizure of the vessel by pirates, the sinking of the vessel during a tsunami or a strong storm, a crash of the vessel against the rocks, a collision with another vessel, running aground, fire, a damage or a failure of mechanical equipment, a shift of the cargo that poses a threat to the vessel stability and other misfortunate events. The cargo faces the following threats (Jarysz-Kamińska 2013, 241–243):

- goods being jettisoned in the case of a sharp tilt of the vessel (a general average)[1]

DOI: 10.1201/9781003330127-11

- the loss of the cargo resulting from a fire, an attack or an extortion of a ransom by pirates for releasing the crew and the vessel
- decay of perishable goods resulting from a failure of cooling equipment
- the quality of goods being worsened as a result of soaking or long storage on the vessel that is detained at a port or unlawfully seized by pirates

Container vessels and their containerized cargo face various risks. According to the data provided by the World Shipping Council (WSC), during the years 2008–2019 around all the seas and oceans of the world, 1,382 containers were lost on average each year as a result of violent storms or human errors (British International Freight Association 2021). The most serious catastrophe in the history of container shipping occurred in 2013 in the Indian Ocean, where the *MOL Comfort*, a vessel with the loading capacity of 8,100 TEU belonging to the Mitsui O.S.K., a Japanese shipowner, sank together with 4,500 containers on board. Under the force of high storm waves and the heavy weight of the cargo, the hull of the vessel broke into half. The stern part of the vessel sank within a few days after the failure. The bow part caught fire during the towage operation and sank after almost a month after the accident. Fortunately, the crew was saved. After the catastrophe, the level of durability defined for vessel hull construction was doubled in comparison to the previous requirements applied by classification associations.

In January 2019, at least 270 containers were lost by the MSC *Zoe* in the North Sea, in the area of the Dutch West Frisian Islands in very difficult weather conditions. After the storm, stranded containers and some goods washed out of the damaged containers could be found along the coastline beaches. In 2020, the *ONE Apus*, a container vessel belonging to the Yang Ming shipowner, lost over 1,800 containers during unfavorable weather conditions over the Pacific Ocean. At the beginning of 2021, also due to difficult weather conditions, the *Maersk Essen*, a Danish container vessel, lost about 750 containers in the Pacific Ocean. Some of the containers that were not taken by high waves were badly damaged. The previously mentioned accidents present a range of problems that can be faced afterward. The comprehensive evaluation of their consequences is presented in Figure 8.1.

The financial loss incurred by the shipowner, the carrier and the consigner/consignee of the cargo is enormous. Jettisoned containers pose a serious threat to other vessels. Such containers usually float just under the water surface, invisible for other navigating ships. In this way, vessels are exposed to very serious collisions or damages. Floating containers or those ones that have sunk to the bottom of the sea come not only as a material loss but also as an ecological threat that becomes especially serious if the containers have been stuffed with hazardous cargo. Assuming that an average value of a container and its cargo is almost USD 25,000, it indicates that the loss resulting from lost containers can reach gigantic amounts.[2] The sinking of a vessel entails even a bigger loss. According to Statista, during the years 2010–2019, 392 commercial vessels sank, including 39 container vessels and 36 ro-ro vessels (Statista 2021d).

To limit negative impact of force majeure events and accidents (risks) in maritime trade, insurance is commonly applied, based on contracts entered into by insurers/insurance companies and insuring parties, usually shipowners and consigners of

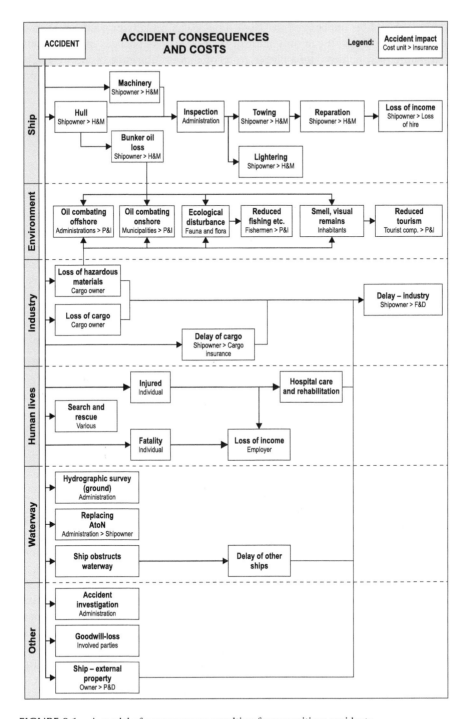

FIGURE 8.1   A model of consequences resulting from maritime accidents.

Source: The authors' own elaboration based on Lundkvist (2010).

Legend: H&M—Hull and Machinery, P&I—Protection and Insurance, F&D—Freight and Delays.

goods. It should be emphasized that no insurance can actually protect parties from unfortunate incidents. It can only mitigate economic consequences of such events, distributing their burden onto a bigger group of entities. Maritime insurance can be divided into three basic groups:

- the hull insurance, which refers to the vessel itself
- the shipowner's civil liability insurance
- additional cargo insurance, which is to protect the interests of the owner of the cargo, not necessarily the interests of the shipowner

8.2   HULL INSURANCE

Similarly to any other seagoing vessels, insurance of container vessels is dedicated to shipowners, also those who operate chartered vessels. This type of insurance is usually referred to as hull insurance or hull and machinery insurance. It is offered mainly by professional insurers. In numerous countries, hull insurance for vessels is most often based on the English Institute Clauses,[3] which were developed in 1982 by the Institute of London Underwriters (ILU) and amended in 2009. There are two types of hull insurance policies:

- Time policy—an insurance policy that is valid for a specified period of time, not longer than one year. The Institute Clauses for hull insurance time policies include *all risks* clauses and other clauses that limit danger, additional clauses that cover more risks, clauses excluding war risk or clauses referring to the particular risks.
- Voyage policy—an insurance policy that is valid for a specified sailing or voyage of a particular vessel. In this type of insurance, a MAR policy is applied together with the Institute Voyage Clauses. The insurance policy agreement usually specifies the date and the port where the voyage starts, transit ports and the destination port.

The H&M insurance covers physical damages caused to the vessel or to the vessel machinery, engines and other equipment as a result of force majeure events or human errors, which are usually covered by the insurance according to the current policy agreed for the particular vessel. Hence, this may include damages caused by accidents that have occurred during the cargo handling operations, explosions, pirate attacks, fires, force majeure events or errors made by the crew of the insured vessel. Additional insurance may cover costs incurred in relation to the rescue operations, emergency towage of the vessel from the place of the accident to the port, the participation of the shipowner in a general average and costs incurred in relation to repairs of other floating objects, wharves or buoys damaged by the vessel (a collision clause) (Kiliński 2018, 15).

The following risks are excluded from the group of risks covered by the H&M insurance: war risk, strike risk, malicious intent risk and nuclear risk. However, shipowners can additionally insure themselves against war and strike risks using special clauses. The H&M insurance may also cover container vessels under construction and ferries. The war-risk policy covers physical damages caused to the vessel as a result of war operations and other events excluded from the basic H&M insurance coverage by the war clauses.[4]

There is a number of offers on the market of maritime hull insurance.[5] The price of a hull insurance policy depends on the size of a vessel, the cargo transported, the length of the shipping route, an insurance option and other factors. Another type of maritime insurance is civil liability insurance.

8.3   CIVIL LIABILITY INSURANCE PERTAINING TO THE OPERATION OF A VESSEL

Shipowners/carriers bear civil liability to their customers, other vessels and ports they call at for various losses that can be caused during operation of seagoing vessels. Sometimes such losses can be of an enormous volume that exceeds any possibilities to be compensated with the use of the carrier's funds. Hence, it is necessary for carriers to get civil liability insurance. However, the shipowner's/carrier's civil liability insurance covers only the risks that result from their direct actions—first of all, erroneous actions. It should be noted that in accordance with the International *Convention* for the *Unification* of Certain Rules of Law Relating to *Bills of Lading* of 1924, amended with the protocols of 1968 and 1979 (the Hague-Visby Rules), shipowners' liability for the cargo they transport is limited. It is defined as a higher amount resulting from the multiplication of SDR 666.7[6] by the number of loading units or SDR 2 per number of kilograms of the gross weight (Marciniak-Neider and Neider, eds. 2014, 699).

Most often, shipowners insure themselves against civil liability risks at shipowners' clubs—at present, about 90% of seagoing vessels are insured at clubs using the P&I or Protection and Indemnity (Kiliński 2018, 17–18) insurance. The main feature of the P&I insurance is the fact that it refers to high risks that are usually difficult for insuring procedures and, considering their large scale, are reluctantly accepted by traditional insurance associations. Shipowners' clubs provide insurance protection to all their members based on the mutual and nonprofit principles.

Shipowners' clubs provide insurance covering accidents that occur to people present on the board of a vessel, damages caused by maritime collisions, pollution caused by vessels to water areas, damages caused as a result of using port facilities, damages that result from providing port services, cargo damages, participation in general averages and fines (Brodecki 2009, 106–127). Shipowners bear civil liability to people present on their vessels: crew members, passengers, people who provide port services on the vessel and other people (visiting family members, guests, survivors, refugees, stowaways). The P&I insurance guarantees the shipowner full coverage and reimbursement of costs incurred in relation to medical treatment, hospitalization, transportation of the injured or the deceased to the country, etc.

Saving people (survivors, refugees) at sea is every seaman's duty. A shipowner may apply to the P&I club for the reimbursement of costs incurred to save such people's lives and their maintenance on the vessel. Stowaways often pose a serious problem to the shipowner and the insurer. Intruders must be provided with minimum food rations and some lodging. The biggest problem, however, comes when stowaways are to be landed at ports because most countries refuse to let them in. The civil liability insurance of the shipowner guarantees the coverage of expenses related to the landing of stowaways in specified countries, such as costs related to proceedings undertaken to establish a stowaway's status, immigration proceedings, repatriation to the country of origin, etc. (Hebel 2005, 89–94).

The shipowner's civil liability insurance also protects them against consequences of vessel collisions that usually entail loss and expenses related to one's own vessel and liability for damages caused to the other ship involved in the collision. Moreover, P&I clubs provide insurance against the risk related to the shipowner's civil liability for polluting the sea with oil. This is usually offered against an additional charge. P&I clubs also cover damages such as the loss or damage of the cargo transported by the vessel, for which the carrier is liable because of the careless loading, stowing, transporting or distributing of the goods.

8.4 CARGO INSURANCE

It should be remembered that the cargo transported by the vessel is not insured against all the risks it can be exposed to during the sailing under the shipowner's/carrier's civil liability insurance. The shipowner is held liable only for the loss caused by their negligence. In accordance with the Hague-Visby Rules, the shipowner is not liable for the following damages, among others (mySped Worldwide Logistics 2021):

- caused by fire, unless it has come as a result of the carrier's actions or the carrier's own fault
- resulting from misfortunate events at sea, such as a damage to the ship, a collision with another floating object, a tilt, having run aground because of a storm, a sea current or a drifting iceberg, etc.
- caused by faulty actions, negligence or errors of the captain, a crew member, a pilot or any other person employed by the carrier in the field of navigation or management of the ship
- caused by force majeure events as a result of natural phenomena of catastrophic nature, a turbulence in social life (war, social and public disorders), sovereign acts that cannot be opposed
- caused as a result of rescue operations undertaken to safe human life or property at sea (participation in a general average)

While taking out a cargo insurance policy, the exporter or the importer may protect themselves against the previously mentioned and other risks that their cargo is exposed to from the moment it is submitted for transportation to the moment it reaches its destination place. The terms of cargo insurance are defined in a policy agreement entered into by the insurer and the insuring party. Having entered into a proper cargo insurance agreement, it is possible to get compensation up to the entire value of the lost or damaged goods. The cargo insurance policy may also cover damages caused by a general average, force majeure events, theft with burglary, robbery or an accident with the participation of a third party.

In accordance with the Incoterms 2020 CIF and CIP, the exporter is responsible for entering into a cargo insurance agreement, and a beneficiary of the insurance (the insured party) is the importer. In considering other trade rules, there is no obligation to insure the cargo. It means that the exporter or the importer, depending on who is held liable for transportation, decides for themselves whether to insure the goods or the cargo for international trade or to resign from such insurance. Usually, in maritime transport, the ICCs (Aquarius Gem Shipping 2021) are applied for cargo insurance, namely the following:

- Institute Cargo Clauses (Clause A)
- Institute Cargo Clauses (Clause B)
- Institute Cargo Clauses (Clause C)
- Institute War Cargo Clauses
- Institute Strike Cargo Clauses 1/1/09
- Institute Theft, Pilferage and Non-Delivery Clause 1/12/82

The Institute Cargo Clause A offers the widest scope of insurance, covering almost all the risks, including protection against costs incurred after the participation in a general average. It excludes risks clearly stated in the exclusion clause—for example, excluding strike or war risks. The insurance rate is calculated with the consideration of the value of goods, types of goods and a shipping route. The insurance is based on a general insurance agreement that includes all the goods that the insuring party wants to transport or to receive within a specified period of time. It can be also based on an individual agreement covering only some goods. A confirmation of an insurance agreement is an insurance policy or an insurance certificate.

In maritime transport, cargo insurance is provided by large insurance companies. A cargo insurance agreement also includes the insuring party's share in the damage, which is stated in percentage or in an amount. It is usually referred to as a conditional franchise. At the conditional franchise, the insurer is relieved from the responsibility for the damage, which does not exceed the amount of the franchise. Usually, the amount is USD 300.[7]

While considering the pragmatics of processes involving maritime container shipping in terms of economic optimization approached in a holistic way, insurance of cargo and vessels comes as the last question. It concludes the considerations of Part 2 dedicated to economic challenges to maritime containerized transport, focused on a holistic approach toward costs in maritime containerized transport.

The problems related to the managing the entire flow of containerized goods also are of great significance for the optimization of processes.

NOTES

1. A general average is an intentional operation undertaken by the captain of the vessel, which involves jettisoning some cargo overboard in order to lighten the vessel and to prevent sinking and saving the vessel, the crew and other part of the cargo. The general average costs are split between the vessel and the cargo, proportionally to their value at the destination place after the sailing is over. If the insurance contract of the insured party covers

general cargo loss, the insured party shall receive reimbursement of the costs incurred in this respect from the insurer.

2. According to the WSC, in 2019, there were 228 million containers transported by sea, with the cargo of the total value of USD 4 billion (British International Freight Association 2021).

3. The Institute Cargo Clauses, the English insurance terms with over 200 years of insurance tradition, are commonly accepted in the field of international cargo insurance. The insurance coverage depends on the set of clauses that have been applied. The Institute Cargo Clauses are of a universal character because they can be applied to the most types of goods and various means of transport. The widest possible insurance coverage, based on the all-risks principle, is provided by the Institute Cargo Clauses (A) 1/1/82. The sets of clauses marked with letters B and C provide limited insurance coverage of the specified risks. In January 2009, the Institute Cargo Clauses (A), (B) and (C) 1.1.09, Institute War Cargo Clauses 1.1.09 and Institute Strike Cargo Clauses 1.1.09 were published. The implemented changes to the previous sets of clauses refer to the modernization of the language, the clarification of some controversial provisions to provide an unambiguous guarantee of protection to the insured parties who do not have any influence on, for example, the choice of a vessel or the changes made to a shipping route.

4. The insurance of a vessel against war risk expires automatically if a war breaks out between any of the five global superpowers (the USA, Great Britain, France, Russia, China), an outbreak of a nuclear war and a vessel requisition.

5. In Poland, the market of hull insurance as well as the entire market of maritime insurance has been dominated by two big insurance companies: Warta and PZU.

6. SDR are special drawing rights, ISO 4217: XDR—an international settlement unit, an account monetary unit of the character of non-cash money that exists only in the form of account records used for deposit bank accounts. It was invented by the International Monetary Fund in 1967 to stabilize the international currency system. In 1970, it was implemented as a preventive tool to fight against a threatening crisis of financial liquidity. At present, it is applied in insurance settlements.

7. In the pragmatics of processes, upon the customer's request, the forwarder may apply for the insurance of the cargo at the moment when they accept the forwarding order. The forwarder insures the cargo only when they receive a clear or a written order to do so. Indicating the value of the cargo does not mean that the forwarder is authorized to arrange a cargo insurance agreement. The forwarder may also (under certain conditions) run liquidation proceedings on behalf of their customer. See more at Aquarius Gem Shipping (2021).

Part 3

Managerial and Operational Challenges to Maritime Containerized Transport

9 The Role of Port Container Terminals in the Maritime Container Turnover

9.1 THE NOTION OF A CONTAINER TERMINAL

In international container turnover, container terminals are of fundamental significance just as much as means of transport used for shipping containers. Container terminals are nodal elements of transport infrastructure. The main task of a container terminal is to provide efficient container handling from one means of transport to another and to move containers around storage yards, with consideration of the optimal use of cargo handling equipment, means of transport and storage surface dedicated to containers. Generally, a container terminal can be defined as a facility where containers are handled from various means of transport in transshipment that can be continued after a particular period of the storage time. In terms of logistics terminology, a container terminal is a designated area with open facilities equipped with proper infrastructure, depending on their functions, which is adjusted to the processes of handling, moving and storing containers (Fertsch, ed. 2006). Hence, a container terminal can be generally defined as a place of a defined surface where cargo that is integrated into ISO containers is stored, sorted and prepared for forwarding (Markusik 2013, 274). Considering the previous statements and for the requirements of this monograph, it is possible to assume that a maritime container terminal includes a space with technical equipment and its IT systems, where container handling operations take place along with other activities related to container turnover. The space includes the following (Bartosiewicz 2020, 130):

- access roads and railway tracks
- parking lots/sidetracks/berths for vessels and stations for other means of transport equipped with container handling facilities
- storage and maneuvering yards for containers
- collection and distribution warehouses
- administration buildings, container repair workshops, washing and cleaning stations for containers

Container terminals are structural elements of ports, and they are included in port functional systems. Operations taking place at a container terminal include processes of handling the particular volume of containers in a configuration depending on the infrastructural and suprastructural capabilities of that terminal, the location of its particular elements and a transport system involving the cooperating branches. The

DOI: 10.1201/9781003330127-13

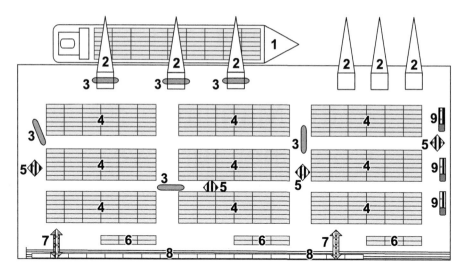

FIGURE 9.1 The operational and organizational structure of a maritime container terminal—a system of a separate container flow.

Source: The authors' own elaboration.

Legend:

1. a container vessel
2. STS (ship-to-shore) cranes
3. yard/terminal tractors for transporting containers to/from the operational range of STS cranes
4. containers stacked on a terminal yard, waiting to be loaded on a vessel or to be transported to the hinterland of the terminal
5. self-propelled container straddle gantry cranes and reach stackers
6. containers waiting for railway transport
7. a crane for unloading and loading containers onto railway cars adjusted for transporting containers
8. a railway sidetrack and a depot of railway cars for transporting containers
9. tractors and semitrailers for transporting containers

operational and organizational structure of a maritime container terminal is presented in Figure 9.1.

In intermodal transport, a network of container terminals comes as a nodal infrastructure located both at the points where land and maritime modes of transport intersect, namely at seaports, and at the points where road and railway modes of transport meet—at land terminals (Waldmann 2016, 198). Maritime and land container terminals generally perform tasks of intermodal terminals. This means that they handle at least two modes of transport.

Considering the criterion related to the location of a terminal in the transport network, there are two basic types of container terminals:

- maritime terminals (that handle road, railway, inland waterway and maritime transport)
- land terminals (that handle road, railway and inland waterway transport)

During intermodal shipping, the goods are transported in one and the same inter-modal transport unit (ITU), so there is no need to strip that unit. A loading unit applied in intermodal transport can be a maritime ISO container, a swap body or a semitrailer.

At large maritime ports, there are usually more than one container terminals. For example, in Rotterdam, there are 26 container terminals and 20 container depots located. Among terminals related to maritime transport, it is possible to distinguish several types of container terminals (Stokłosa, Cisowski and Erd 2014, 5933):

- Deepwater terminals—usually located at seaport entrances, along impor-tant maritime shipping routes; they can handle the largest container vessels in the world.
- Hub terminals—usually located at hub ports, from where some containers are transported by feeder vessels to smaller ports; containers are also trans-ported from smaller ports to hub ports in the same way.
- Feeder terminals, where containers are transported from hub ports or other ports; containers are also transported from feeder terminals to hub ports or other destination ports located on the same continent.
- Land container terminals or dry ports.
- Container depot terminals.

Considering concepts of integrated intermodal transport, the development of mar-itime ports—supported by proper legislation (at the national, regional and European levels)—fosters extended accessibility of infrastructure. This, in turn, results in the fact that a number of optimizing solutions are applied, such as shipowner container depots or land dry ports.

9.2 DRY PORTS

A dry port is an intermodal terminal located in the hinterland of one or several sea-ports, which is linked with those ports by regular railway and/or road connections and which is dedicated to handle containers, offering services to forwarding com-panies and carriers who operate at seaports (Grulkowski and Zariczny 2012, 188). Dry ports are constructed to eliminate bottlenecks at seaports that often result from the lack of storage place to keep full and empty containers for long periods of time. Dry ports operate partially as port terminals, largely eliminating transport conges-tions often observed in the vicinity of seaports. Depending on their location and the range of services they provide, dry ports can be divided into three types (Neider and Marciniak-Neider 1997):

- satellite terminals
- logistics center terminals
- cargo handling terminals

The main task of satellite terminals is to limit congestion in the nearest vicinity of a particular port. Apart from container storage, a satellite terminal usually offers services such as stuffing and stripping containers, customs clearance of goods in trade with the countries from the outside of the EU customs territory, cleaning and repairing containers and others. Container terminals that operate as logistics centers are located on the outskirts of large cities, in the area or in the vicinity of other logistics centers, technology parks or duty-free zones. They are connected by fast communication lines with one or more seaports. At dry ports, containers are handled, and various logistics services are provided, such as storage, forwarding, etc. Cargo handling dry ports are located at major communication nodes, where containers are transported from and to ports by shuttle trains and from where containers are delivered to customers, directly to their destination addresses. Considering the process of container handling in such a way, container depots perform a slightly different role.

9.3   CONTAINER DEPOTS

A container depot or container storage facility is a place where empty containers are kept. Container depots can be divided into the following (Kurek and Ambroziak 2017, 150–153):

- maritime depots that are located in the vicinity of seaports or at seaports
- land depots that are most often located in the vicinity of large railway and road hubs

Apart from short-term and long-term storage of empty containers, container depots also offer some additional services related to containers and to the cargo transported inside them: receiving and forwarding cargo, stuffing and stripping containers, consolidating cargo, customs clearance of goods, transporting cargo by railway or by trucks to and from the port, washing, fixing, maintenance, modification and repairs of containers. Container depots can be the property of the state/city or of a private entity, or they can be publicly-privately owned. In Europe, North America and East Africa, container depots most frequently belong to container terminal operators, shipowners or railway companies. In China or in India, container depots are usually state-owned facilities. Container depots allow land carriers and shipowners to use their transport capabilities in a more efficient way. They also come as an important factor to the development of intermodal transport in the world. A dense network of container depots also allows carriers to increase their customer bases and to respond to changes observed in the market of container shipping in a more flexible way. Table 9.1 presents the largest container depots in the world, according to the data provided by an internet platform.

Considering such a wide range of potential functionalities and operational activities performed at container terminals, it is important to make an attempt at providing the characteristics of a maritime container terminal (based on its specificity).

TABLE 9.1

The Largest Container Depots in the World

Port	Country	Depot	Storage capability
Shanghai	China	Donghwa Container	53,000 TEU
Ningo	China	Ningbo Victory Cntr Co., Ltd.	30,000 TEU
Antwerp	Belgium	DR Depot	23,000 TEU
Rotterdam	Netherlands	DR Depot	20,000 TEU
Hamburg	Germany	Progeco Deutschland GmbH-Ellerholzdamm	No data provided
Genoa	Italy	Gruppo Spinelli	10 thousand TEU
Qingdao	China	Ocean & Great Asia Logistics Co., Ltd	No data provided
Klang	Malaysia	Eng Kong Holdings Pte Ltd.	No data provided
			No data provided
Singapore	Singapore	Ningbo Victory Cntr Co., Ltd.	No data provided
Nhava Sheve	India	Bay Container Terminal Pvt	No data provided

Source: Data from X Change (2021).

9.4 THE CHARACTERISTICS OF MARITIME CONTAINER TERMINALS

As an important link in land–sea transport chains, a seaport is a cargo concentration (consolidation) center in export for consigners and a cargo deconcentration (deconsolidation) center in import for consignees (Misztal, ed. 2010, 13). The processes are implemented with the use of homogeneous international transport (maritime transport) and diversified national transport systems (applying means of transport of various transport modes) that fit the logistics concept of a hub and spokes (Miler 2016a, 44).

Among all ports of such a character, container ports present the highest concentration of logistics services and processes that take place there with the highest intensity as a result of operations undertaken by logistics groups (so called logistics mega-operators). Logistics groups and mega-carriers have already started to intensify the organization of their activities around the network of container terminals. Hence, terminals have become a driving force for the cooperation among entities operating at ports, which is to optimize logistics processes implemented also in a broadly understood hinterland. In this way, organization of ports has been transformed into "a terminal community" (Tubielewicz and Miler 2014, 140). The characteristics of such relations is presented in Figure 9.2.

The field of external—or in other words, vertical—adjustments provides a convenient system of transport relations with the hinterland and maritime foreland. The relations with the foreland involve establishing the main networks of shipping connections. The hinterland includes, first of all, road, railway and pipeline connections and also—more and more often—inland waterway connections.[1]

Maritime container terminals, also referred to as port container facilities, include port wharves with one or two berths for handling vessels, cargo handling and storage yards, gates for trucks, railway sidetracks for handling cargo, collection and

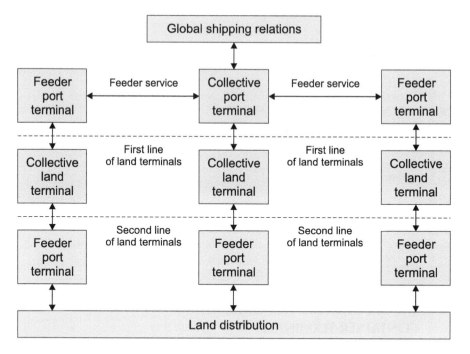

FIGURE 9.2   A stage of seaport terminalization.

Source: The authors' own elaboration based on Vanroye and van Mol (2008, 68).

Legend: intermodal connections (road, railway, inland waterway transport)

distribution warehouses, a coordination and control center, container washing and cleaning stations, repair workshops and technical cargo handling facilities.[2] The structure and organization of processes taking place at a maritime container terminal are presented in Figure 9.3.

The length of wharves and depth of port basins determine the number of berths and the size of vessels that are handled. Considering handling ro-ro vessels with the vertical loading systems, wharves must be equipped with proper loading ramps. Cargo handling and storage yards are usually adjacent to the wharves. They are dedicated to short-time storage of containers, and they are divided into sectors where lanes for maneuvering vehicles are designated. The surface of a storage yard must be reinforced considering the heavy weight of containers that are stored there or moved around the yard with the use of maneuvering vehicles. In the close vicinity of wharves, cargo collection and distribution warehouses are usually located where containers are stuffed and stripped. Warehouses are equipped with high numbers of gates and loading ramps. Sometimes a railway track can be incorporated into the interior of a warehouse. An important element of such a container facility is a gate for trucks, where containers and all the necessary documents are received and dispatched. It is also a place where all instructions referring to the storage of containers are given. The gate is equipped with special scales for weighing trucks

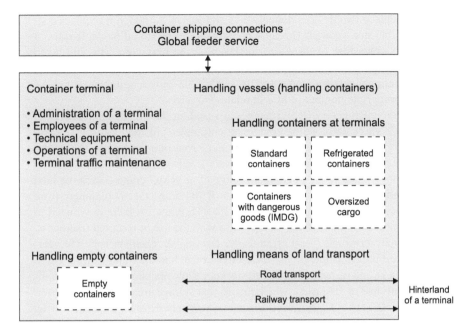

FIGURE 9.3 The scheme of processes taking place at a maritime container terminal.

Source: The authors' own elaboration based on Green Efforts (2021).

with containers. The scales can be also located in the proximity of the gate. Another important element of a port terminal is its coordination tower center, from where the entire terminal can be observed.

Contemporary maritime container facilities are characterized by high amounts of equipment, the character and size of which depend on the technology applied to handle cargo by the size of terminal area and by the volume of containers handled there. The landscape of a port terminal is usually dominated by robust self-propelled STS (ship-to-shore) gantry cranes traveling along special rails, with their long booms visible from the sea and land sides. Considering operational processes, it is possible to distinguish four main technological solutions that are applied at container terminals (Marciniak-Neider and Neider, eds. 2014, 340–341):

- Containers are unloaded from the vessel onto semitrailers with the use of STS cranes; then they are taken by terminal tractors to the storage yard. At the storage yard, containers are unloaded from semitrailers by RTG (rubber tire gantry) cranes and stacked. Containers are dispatched from the storage yard in the following way: they are taken out of the stack by an RTG crane and loaded onto a truck or a railway car. The previously mentioned solution is frequently applied, considering the efficient use of the storage area and high cargo handling efficiency as well.
- Straddle carrier (SC) tractors are used for moving cargo around storage yards and loading cargo onto trucks or railway cars. Hence, SC tractors

can replace two devices, namely TTs and RTGs. Although it requires considerable storage areas (because containers can be stored in single rows of stacks consisting of three or four layers at the maximum), the discussed technology is often applied in developing terminals as it does not require any permanent reinforcement of the surface; the only requirement is to have an appropriate number of SCs in relation to STSs.

- Rail mounted gantry (RMG) cranes are used for moving containers around storage yards and loading containers onto means of road transport. RMG cranes allow operators to stack containers even into ten layers. The operational range of a RMG crane also covers a railway sidetrack and a loading lane for trucks. If the operational range of an RMG crane covers the land operational range of an STS crane, an RMG crane takes containers from the point where they are blocked by an STS crane, so there is no need to transport containers to the storage yard with the use of terminal tractors.

- At small terminals (most often feeder terminals), reach stacker (RS) vehicles are used for operations related to storage, handling and loading cargo on trucks and railway cars. Reach stackers allow operators to stack containers into five or six layers, which means that storage area is used in a better, more efficient way. However, reach stackers require a lot of maneuvering space.

Considering the pragmatics of their operation, maritime container terminals are managed by specialized entities often referred to as operators of maritime container terminals.

9.5 THE SELECTED GLOBAL OPERATORS OF MARITIME CONTAINER TERMINALS

Due to the implementation of uniform standards in management, facilities (infra- and suprastructure), equipment and computerization (leading to autonomation and robotization), operators of maritime container terminals are able to achieve the economies of scale (both in terms of economy and operation) that contribute to the improvement in process optimization and the achievement of more efficient terminal functioning (also in terms of competitiveness). The largest global operators of maritime container terminals are presented in Table 9.2.

PSA INTERNATIONAL

The biggest operator of maritime container terminals is PSA International, with its headquarters based in Singapore. Today its second headquarters is based in Antwerp. The company was established in 1964 as the Port of Singapore Authority (PSA)—that is, as an institution to manage the port of Singapore. In the 1970s, the PSA launched its pioneer project of constructing a container port in Singapore, which handled the first container vessel in 1972. The container turnover in Singapore was developing in a very dynamic way: in 1982, the volume of containers handled by PSA exceeded the level of one million TEU, and in 1990, that level was increased to

TABLE 9.2

5 Top Operators of Maritime Container Terminals in 2019 (Stated in Million TEU)

No.	Operator	Headquarters	Cargo handling volume
1	PSA International	Singapore	60.4
2	China Cosco Shipping	Shanghai	48.6
3	APM Terminals	Hague	46.8
4	Hutchison Ports	Hong Kong	45.7
5	DP World	Dubai	44.3

Source: Data from Statista (2021f, 32).

TABLE 9.3

Container Handling Operations at the Terminals of PSA International during the Years 2009–2019

Year	2009	2010	2011	2012	2013	2014	2015	2016	2017	2018	2019
Million TEU	47.4	54.1	57.1	60.1	61.8	65.4	64.1	67.6	74.2	81.0	85.2

Source: Data from Statista (2020h).

5 million TEU. This made PSA the largest operator of port terminals in the world. In 1996, the regulation tasks of the company at the Singaporean port were taken over by a newly established Maritime and Port Office in Singapore and PSA focused its operation on handling container terminals. Also in 1996, PSA started its expansion onto foreign markets, engaging its capital into the port of Dalian in China. at present, PSA International is an enormous investment holding company for the group of PSA entities in the world. In terms of the capital, it remains under control of Temasek Holdings, the Singaporean state investment fund. PSA International runs its business operations in 26 countries, at 50 ports, handling 60 maritime, railway and inland waterway container terminals. The terminals are located in Asia, the Middle East, Europe and both Americas.

In 2019, the terminals belonging exclusively to PSA International or operating as joint venture companies established with their business partners employed 38,000 people. The terminals were equipped with 1,800 modern cranes and other container handling equipment. A number of 85.2 million TEU were handled at those terminals, including 36.9 million TEU at the terminal in Singapore (Table 9.3).

The operational entities of PSA International have been divided into five business regions: Southeast Asia, Middle East and South Asia, Northeast Asia, Europe and Mediterranean Sea and America. Each region is responsible for its business performance. Regional directors report to general group director, who is assisted by a small team of top managerial staff of the Corporate Centre in Singapore. PSA plays an important role in global supply chains and in the development of modern technologies in the field of container turnover, especially in the IT sector (PSA International 2021).

In 2019, PSA International, the Polish Development Fund and the Global Infrastructure Fund, an Australian investment fund, jointly bought 100% of shares in the Deepwater Container Terminal Gdańsk (DCT Gdańsk). In this way, PSA International launched its expansion in Central and Eastern Europe. The Singaporean company has already announced a project of increasing the cargo handling capabilities of the DTC Gdańsk from the current level of 2.2 million TEU up to 7 million TEU in the future.

Cosco Shipping Ports (CSP)

The second largest port operator of container terminals is the Chinese Cosco Shipping Ports (CSP), which since 2016 has been operating under the China Ocean Shipping Company (COSCO Group). The company headquarters are based in Hong Kong; however, it is officially registered in Bermuda. Business operations of CSP are focused on port container terminals located in China and abroad. The company also provides other numerous services that are related to maritime transport: container production and leasing, ship repairs, forwarding, freighting, etc.

According to the data provided at the end of 2019, CSP ran its business operation at 36 ports, where it managed and operated 290 berths for vessels, including 197 berths dedicated to handle container vessels, which represented the total cargo handling capabilities of 115 million TEU annually. Apart from China, the company is an operator of container terminals located in Singapore, Japan, South Korea and numerous ports, such as Antwerp, Valencia, Bilbao, Vado Reefer (Italy), Piraeus (Greece), Kumport (Turkey), Seattle (USA), Chancy (Peru), Dubai and Port of Suez. In 2019, the volume of cargo handling operations at the CSP container terminals reached the level of 130 million TEU, out of which 29.2 million TEU (that is, 22.5% of the total cargo handling volume of the entire company) accounted for the turnover recorded by the foreign container terminals. At the terminals remaining under the control of CSP, there were 26.6 million TEU handled, and 103 million TEU were handled at the terminals with the CSP's capital minority share (Cosco Shipping Holdings Co. Ltd. 2020, 26). The foreign CSP terminals are presented in Table 9.4.

APM Terminals

Established in 2001 with its headquarters in the Hague, APM Terminals is also included in the group of the biggest global operators of maritime container terminals. It belongs to Maersk, a Danish maritime conglomerate. APM Terminals runs its operations at 74 maritime container terminals located in 58 countries on five continents: Europe, Asia, Africa, North and South Americas. The company employs over 20,000 workers. APM Terminals operates two container terminals at each of the following ports: Rotterdam, Vado, Abidjan, Tangier, Laem Chabang and Tianjin. Furthermore, the company operates four terminals in St. Petersburg and five terminals in Qingdao. In 2019, APM Terminals was the controlling shareholder in the majority of the previously mentioned 74 terminals; the ownership of the remaining terminals is shared with the company's business partners in the form of the joint

TABLE 9.4

The Foreign Container Terminals of CSP (2019)

Name of the terminal	The capital share of CSP, stated in %	The number of wharves	Cargo handling capabilities (stated in million TEU)
Piraeus Terminal	100	8	6.2
Suez Canal Terminal	20	8	5.0
Kumport Terminal	26	6	2.1
Zeebrugge Terminal	85	3	1.3
Antwerp Terminal	20	4	3.7
COSCO-PSA Terminal Singapore	49	5	4.85
Busan Terminal	4.9	8	4.0
Seattle Terminal	13.3	2	0.4
Euromax Terminal	35	5	3.2
Abu Dhabi Terminal	90	3	2.5
Vado Reefer Terminal	40	2	0.25
Valencia Terminal	51	6	4.1
Bilbao Terminal	39.8	3	1.0
Chancay Terminal 60%	60	2	1.0

Source: Data from Cosco Shipping Holdings Co. Ltd. (2020, 37).

TABLE 9.5

Container Handling Operations at the Terminals Handled by APM Terminals during the Years 2016–2017

Region	2016 in million TEU	2016 in %	2017 in million TEU	2017 in %
America	6.4	17.2	4.4	11.1
Europe, Russia, the Baltic region	11.8	31.6	12.7	32.0
Asia	12.5	33.5	13.6	34.3
Africa and the Middle East	6.6	17.7	7.0	17.6
Total	37.3	100.0	39.7	100.0

Source: Data from APM Terminals (2017).

venture enterprises (APM Terminals 2021). Apart from operating container terminals, APM Terminals provides numerous services related to the field of maritime transport, such as leasing containers, cleaning and repairing containers, organizing supply chains, stuffing containers, storing goods, etc.

APM Terminals runs its business operations in four geographical regions, namely in America, Europe (including Russia and the Baltic region), Asia and the Middle East. As stated in the available data referring to the years 2016–2017, the highest share in container handling operations, stated in TEU, was recorded in Asia and Europe, as seen in Table 9.5 (Statista 2021f).

According to Statista (Statista 2021f, 32), in 2018, cargo handling volume at the terminals operated by APM Terminals reached the level of 46.8 million TEU.

HUTCHISON PORTS

Another leading company on the list of the biggest global operators of port terminals is Hutchison Ports, with its headquarters based in Hong Kong. It is officially registered the Virgin Islands and is a subsidiary of CK Hutchison Holdings. The company started its operation in the foreign markets in 1991. Over the years, Hutchison has diversified its business operations, providing services in the fields of logistics, ship repair, handling passenger vessels, airports, railway terminals distribution centers and other activities.

In 2019, Hutchison Ports was an operator of 52 container terminals with 290 berths for vessels in 27 countries (including ports in Barcelona, Busan and Buenos Aires) on all the continents. The companies belonging to Hutchison employ 30,000 people, and the capacity of containers handled in 2019 reached the level of 86 million TEU.

DUBAI PORTS WORLD (DPW)

Another operator from the list of the largest global operators of maritime and land container terminals is Dubai Ports World (DPW), with its headquarters based in Dubai. DPW handles almost 10% of the global container trade, and it is an integral part of a big Dubai World Holding Group, controlled by the government of the United Arab Emirates. DPW was established in 2005 as a logistics company as a result of a merger of Dubai Ports International and Dubai Ports Authority. Initially, the company focused its business operation on logistic cargo handling, but in the course of time, it started to take over various companies related not only to maritime trade and to diversify its business activities. At present, apart from its core business operation—namely managing maritime container terminals—DPW also provides other maritime services (port and vessel management, solutions in the field of environmental protection at ports, counteracting maritime piracy, etc.), management of duty-free zones and services in the field related to optimization and security of supply chains. DPW owns over 80 container terminals, and according to Statista in 2018, the volume of cargo handling operations performed only at the terminals fully controlled by DP World reached the level of 44.2 million TEU (Statista 2021c). On its websites, DP World informs that in 2019, 71 million TEU were handled at the terminals with the company's shareholding (not only controlled by the company). A number of 14.1 million TEU of the previously mentioned volume were handled at the terminals located at the Port of Jebel Ali (DP World 2021a). During the first half of 2020, DP World ran its operations at 127 business units (mostly Asian subsidiaries) located in 51 countries on six continents, where 56,000 people were employed (DP World 2021b).

INTERNATIONAL CONTAINER TERMINAL SERVICES, INC. (ICTSI)

International Container Terminal Services, Inc., a Philippine business organization, with its headquarters based in Manila, is another global operator of container terminals. It was established in 1987 by a Philippine businessman, Mr. Enrique K.

TABLE 9.6

Container Handling Operations at the Terminals Managed by ICTSI during the Years 2010–2019 (Stated in Million TEU)

Year	2010	2011	2012	2013	2014	2015	2016	2017	2018	2019
Million TEU	4.2	5.2	5.6	6.3	7.4	7.8	8.7	9.2	9.7	10.2

Source: Data from ICTSI Annual Report (2020) (the years 2010–2019).

Razon, whose family has been managing Philippine ports for three generations as a result of an invitation of the Philippine government to a tender for the management of Manila International Container Terminal (MICT). After it had got some experience and reinforced its capital basis, in 1994 the company launched a program of intensive expansion onto the national and international markets of port services. It started to take over container terminals all over the world, especially at countries the governments of which intended to privatize those terminals and their management. Now, the company invests a lot into port structures, equipment of container terminals, IT and training dedicated to the employees of the terminals belonging to the company. At the end of the second decade of the 21st century, ICTSI employed approximately 7,000 employees. In the company portfolio, there are 32 container terminals located in 19 countries, including Poland. Among these terminals, there is Victoria International Container Terminal in Melbourne—the first fully automated maritime container terminal in the world. The number of terminals operated by the company will surely grow because ICTSI continues to pursue the policy of acquiring new contracts for managing terminals on all the continents. In 2019, the capacity of containers handled at the port terminals of ICTSI was increased from 4.2 million TEU in 2010 to 10.2 million TEU (Table 9.6).

ICTSI has divided its business operations related to port terminals into three geographical regions:

- Asia-Pacific: ports of the Philippines, Karachi (Pakistan), Yantai (China), Jakarta and South Sulawesi (Indonesia), Port of Moresby and Lae (New Guinea), Melbourne (Australia)
- America: ports of Manzanillo (Mexico), Cortes (Honduras), Buenaventura (Colombia), Pernambuco and Rio de Janeiro (Brazil), Guayaquil (Ecuador), Buenos Aires (Argentina)
- Europe–the Middle East–Africa: ports of Gdynia, Rijeka (Croatia), Batumi/Adjara (Georgia), Umm Qasr (Iraq), Toamasina (Madagascar), Matadi (the Democratic Republic of the Congo), as seen in Table 9.7

As indicated in the table presented previously, the operation of ICTSI is mainly focused on Asia and the Pacific region, whereas Europe plays a minor role in the company's strategy.

TABLE 9.7

The Geographical Structure of the Container Terminals Remaining at the Disposal of the ICTSI in 2019, by the Volume of Cargo Handling Operations

Region	Number of terminals	Million TEU	Share stated in %
Asia-Pacific	19	5.4	53
America	7	3.0	29
Europe–the Middle East–Africa	6	1.8	18
Total	32	10.2	100

Source: Data from ICTSI Annual Report (2020).

EUROKAI

The leading operator of container terminals in Europe is a German-Italian company, Eurokai, with its headquarters based in Bremen, which has been running its business operations since 1999. The company operates in Germany, Italy, Portugal, Morocco and Russia. Its core activities include organizing handling operations and handling containers at seaports with the use of modern technical facilities, including the IT assistance. The company operates container terminals in Bremerhaven, Hamburg, Wilhelmshaven, Ravenna, La Spezia, Salerno, Lisbon, Tangier, Limassol and Ust-Luga. In 2019, 11.65 million TEU were handled at the container terminals of Eurokai (Table 9.8). Additionally, Eurokai—as any other operator of container terminals—provides a number of additional services, such as intermodal transport of cargo from land economic centers to seaports, and the other way round, storing empty containers, repairing, stuffing and stripping containers, maintenance and repairing port facilities and handling components for wind power plants.

During the years 2018–2019, the container turnover of the Eurokai Group was lower than during the years 2016–2017 because the container terminal in the Italian port of Gioia Tauro was sold and not included into the statistical data. In 2019, the Eurogate container terminals in Germany received containers delivered by intermodal transport, which represented the capacity of 680,000 TEU and 313,000 TEU in Italy and 102,000 TEU in Portugal (Eurokai Hauptversammlung 2021).

TERMINAL INVESTMENT LIMITED (TIL)

A company named Swiss Terminal Investment Limited (TIL), with its headquarters based in Geneva, also belongs to the group of large operators of container terminals. It was established in 2000 to secure vessel berths and ensure the efficient handling of vessels that belonged to one of the largest shipowners in the world, namely to the Mediterranean Shipping Company (MSC) at container terminals. In 2019, TIL owned 40 container terminals in 27 countries. The company's most important

TABLE 9.8

Container Handling Operations at Eurokai Terminals during the Years 2016–2019 (Stated in Million TEU)

Terminal	2016	2017	2018	2019
Germany, including:	8.23	7.78	7.76	7.60
Hamburg	2.26	1.69	1.63	2.09
Bremerhaven	5.49	5.54	5.47	4.87
Wilhelmshaven	0.48	0.55	0.65	0.64
Italy, including:	5.01	4.64	2.07	1.91
Gioia Tauro	2.75	2.39	no data provided	no data provided
Calgari	0.64	0.40	0.20	0.05
La Spezia	1.14	1.34	1.35	1.30
Salerno	0.27	0.31	0.33	0.37
Ravenna	0.20	0.18	0.18	0.18
Others, including:	1.36	1.99	1.97	2.14
Tangier	1.13	1.38	1.37	1.53
Limassol	no data provided	0.34	0.39	0.41
Lisbon	0.15	0.19	0.13	0.14
Ust-Luga	0.08	0.07	0.06	0.06
Total	14.61	14.41	11.68	11.65

Source: Data from Eurokai (2017); Eurokai (2019).

customer is the MSC shipping line. The operator takes over container terminals located on important maritime routes, usually by entering into 50/50 joint venture enterprises with their previous owners. Next, TIL modernizes, extends and operates such terminals. TIL terminals are located in geographically significant regions of the world, where high container turnover can be observed. The company operates its container terminals in five geographical regions (TIL Terminals 2021):

- North Europe—terminals are located in Antwerp, Rotterdam, Liverpool, Le Havre, Bremerhaven, Klaipeda and St. Petersburg
- South Europe and Africa—terminals are located in Marseille, Ashdod (Israel), Genoa, Gioia Tauro (Italy), San Pedro (Ivory Coast), Sines (Portugal), Las Palmas and Valencia (Spain), Asya Port, Iskenderun and Istanbul (Turkey)
- North America—terminals are located in Montreal and some ports of the USA, such as Everglades, Freeport, Houston, Long Beach, New Orleans, Newark, Seattle
- South and Central America—terminals are located in ports of Buenos Aires, Bahama, Rio de Janeiro, Navegantes and Santos (Brasil), Rodman (Panama), Callo (Peru)
- Asia—terminals are located in ports of Abu Dabi (UAE), Ningho (China), Mundra (India), Red Sea (Saudi Arabia), Umm Qasr (Iraq), Singapore

According to Drewry Maritime Research, the volume of containers handled at TIL terminals in 2019 reached the level of 28,800 TEU (Lloyd's 2021).

The container terminals of the TIL Group come as important links connecting global container shipping lines with economic centers in the hinterland, where goods transported by vessels come from and where cargo brought from overseas countries is delivered. Shipping containers between port terminals and economic centers located in the hinterland is implemented by road, railway and inland waterway modes of transport. Hence, shipping containers from manufacturers to consumers is implemented by intermodal transport, and its efficient functioning requires adequate coordination of all modes of transport involved. The container terminals of the TIL Group are responsible for this task.

Moreover, container terminals function as transmission hubs in the networks of container shipping lines. There containers are handled between vessels operating along the main lines and between vessels operating along the main lines and feeder lines. In this way, shipping companies can expand their operation onto various markets and improve operation of container vessels. Apart from the previously mentioned services, the container terminals of the TIL Group provide additional services, such as stuffing and stripping containers, storage of goods, repairs and storage of empty containers, securing port wharves and others.

EVERGREEN MARINE CORPORATION (EMC)

Evergreen Marine Corporation (EMC) is another good example of an operator of container terminals that belong to global shipping companies. The company was established in 1968 as a Taiwanese maritime corporation, with its headquarters based in Taipei. The company offers a wide range of sea-related services and services in the field of maritime transport. As an operator of container terminals, it is one of the ten largest entities in the world that manage and operate container terminals. Business operations of EMC also involve producing, repairing, storing and stuffing containers, constructing vessels, forwarding and real estate management. EMC operates two main types of terminals (Lloyd's 2021):

- four large container hubs—two in Taiwan (Taichung and Kaohsiung), in Colon (Panama) and Taranto (Italy)
- other container terminals located in the USA, Asia and Africa

In 2019, the volume of container handling at ECM terminals reached the level of 9.5 million TEU.

PACIFIC INTERNATIONAL LINES (PIL)

The group of the ten largest operators of the global fleet of container vessels includes a Singaporean maritime corporation established in 1967, Pacific International Lines (PIL). At the beginning, the operation of the company was related to short-sea shipping, but at the turn of the 1980s and 1990s, it was transformed into a global operator of the container vessel fleet. The company employs over 9,000 people. At the beginning of 2020, it operated 91 container vessels of the capacity of 279,000 TEU, including 34 chartered vessels of the capacity of 153,000 TEU (AXSMarine 2021). PIL focuses its operation on container shipping among ports of Asia, Africa, the Middle East, Latin America, Australia and Oceania. PIL container vessels call at ports located in 100 countries. These are both hubs and small ports located in the Bay of Bengal, on the coast of India, on the Red Sea, on the western coasts of Africa and on the islands of Oceania, where containers are delivered by feeder vessels. Apart from container shipping, the Pacific International Lines also provides bulk cargo shipping and logistics services. The company is also the owner of maritime agencies; it offers services related to repairs and disposal of containers. The company holds shares in Chinese container-manufacturing plants through the Singamas company, and it also manages some large container depots in Singapore, Indonesia, Bangladesh, New Zealand, Fiji, Egypt and Tanzania (Pacific International Lines 2021).

As presented in this chapter, the findings clearly indicate the fact that in international maritime container turnover, the key role is performed by container terminals (and their operators) next to the means applied to transport containers. Operators are responsible for the elements of competitiveness characterizing particular terminals. These elements also include organizational and technical conditions required for the operation of port container terminals as well as other numerous components (determinants).

NOTES

1. Rydzkowski and Wojewódzka-Król (2006, 194–200).
2. Krasucki and Neider (1986, 87).

10 Organizational and Technical Conditions Underlying Operation of Port Container Terminals

10.1 MODELS OF THE FUNCTIONING OF PORT CONTAINER TERMINALS

Considering the fact that port container terminals operate under the regimes of high competitiveness, pressure of process optimization and drastically strict requirements related to lowering anthropopressure, it is possible to identify some activities undertaken by terminal operators to use those elements in order to achieve the competitive edge. Some excellent examples of such activities are functional models of the following:

- a modernized conventional terminal—based on the model of standard (conventional, horizontal) operational characteristics, e.g., Luka Koper, a maritime container terminal in Slovenia
- a terminal with the vertical structure of processes implemented to optimize them, e.g., Busan New Container Terminal (BNCT) in Busan (Korea)
- a low-emission/zero-emission terminal, a terminal in the port of Hamburg (Germany)—HHLA Container Terminal Altenwerder (CTA)—the first terminal in the world that has been certified under the *Voluntary Emission Reduction* (VER) program as neutral for the climate
- an intermodal improved-port/ship-interface (IPSI) terminal, based on intermodality in the field of water transport (maritime and inland waterway ro-ro transport)

Each of the previously mentioned terminals is characterized by some particular features that determine the ways they operate and which are the reasons for some significant changes to the model of architecture of processes applied at the terminals (the process layout).

The Luka Koper maritime container terminal (Slovenia) comes as an example of a modernized container terminal, which fits a conventional (standard and commonly applied) process model. The Luka Koper container terminal is located in the area of the port basin I, quay 7 of the maritime Port of Koper, and it has been systematically developed. The initial length of its wharves, which was 596m in 2010 (divided into four berths of different length ranging between 100m–200m), has been so far significantly extended. Initially, there were eight STS cranes (4 × Postpanamax and

DOI: 10.1201/9781003330127-14

4 × Panamax) to handle container vessels at the terminal; however, at present (2022), there are 11 STS cranes (4 × super Postpanamax, 4 × Postpanamax and 3 × Panamax) at the Luka Koper terminal. SuperPostpanamax and Postpanamax vessel (mother vessel) handling operations usually take place at the western side of the quay, considering the required depth parameters. The location of the berths and the operational range of the STS cranes at the Luka Koper container terminal are presented in Figure 10.1.

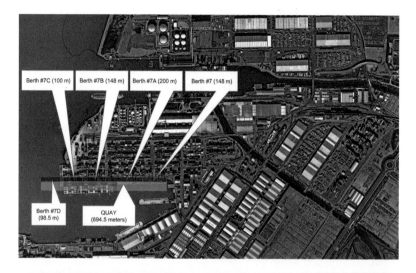

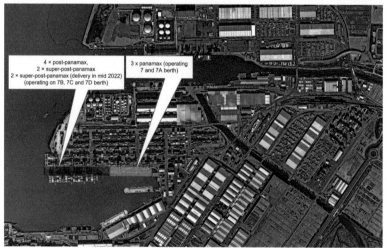

FIGURE 10.1   The wharf at the Luka Koper container terminal, including the operational range of the STS cranes.

Source: Reprinted with permission of Šik Sebastjan Port of Luka Koper Head of PR Department.

Legend: Left, berths at the wharf; right, the operational range of the STS cranes.

From the quay side, behind the operational range of the STS cranes and typically for a container terminal, there are some areas dedicated to the storage of full containers (import/export) and empty containers (import/export). In the area adjacent to the operational range of the STS cranes, further in the yard, there is usually an area where reefer containers are kept, marked as "PTI passed" (next to the PTI zone—a pre-trip inspection zone). There is also a container depot in the further vicinity of the main operational area. The container storage system applied in the yard is presented in Figure 10.2.

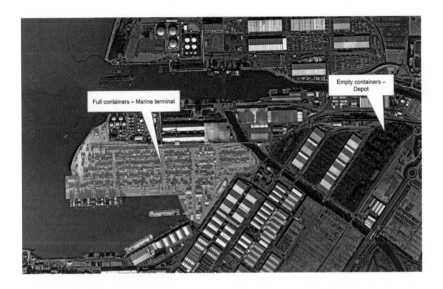

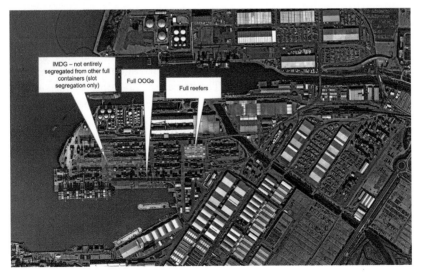

FIGURE 10.2   The container storage system at the Luka Koper container terminal.

Source: Reprinted with permission of Šik Sebastjan, Port of Luka Koper Head of PR Department.

FIGURE 10.3   The infrastructure providing access to the Luka Koper maritime container terminal.

Source: Reprinted with permission from Šik Sebastjan, Port of Luka Koper Head of PR Department.

Legend: Blue, the road system; red, the railway sidetracks.

Apart from areas dedicated to container storage, the terminal yard is also equipped with the infrastructure providing access to the means of road transport (truck and trailers) and standard gauge trains—there are three sidetracks. The access to the container terminal, the gate and railway tracks are presented in Figure 10.3.

As presented previously, the location of the particular elements of the infrastructure providing access to the container terminal and the conventional system of locating operational zones determine the course of container handling processes, where it is possible to distinguish the following:

- Handling containers in export:
  - Loaded on semitrailers, containers pass through the main gate; according to their notifications (considering the current cut-off time), they are visually checked for damage. The place for unloading them is designated, and RTG cranes unload the containers from the semitrailers and stack them into layers.
  - Containers are delivered by railway and unloaded by RTG cranes or reach stackers onto terminal tractors, which take them to designated places (particular stacks on the container export yard), where containers are unloaded by RTG cranes.
  - When needed, exported containers are taken off the stacks by RTG cranes, and in accordance with the specific loading list, they are taken by terminal tractors to the maneuvering area on the quay and loaded onto a particular vessel by STS cranes in the appropriate order.
- Handling containers in import:
  - Containers are unloaded from a vessel by STS cranes and transported by terminal tractors to the place designated on the import container yard for storage (handled there by RTG cranes).

- Containers that are imported by railway cars undergo procedures that are reverse to export procedures (handled by RTG cranes and reach stackers).
- Containers are picked up by the means of road transport directly from the storage yard (containers are handled by RTG cranes).
- Additional activities related to some selected containers (usually, these are containers undergoing customs, veterinary, phytosanitary or technical procedures) with specific annotations (or seals) in the system of terminal management and monitoring.

A more complicated process takes place when containers are loaded and unloaded onto or from a vessel by STS cranes. Considering the fact that containers are stowed on the vessel with the use of twist locks and lashes, the process must be handled by employees who are responsible for those types of cargo securing methods (and for the entire stowing process and STS crane operations). The lashing system is taken off on the vessel; twist locks are often taken off or fixed on during transportation of containers on terminal tractors (which are equipped with adequate pockets facilitating the whole process). Stowage operations are divided between the members of stowing teams: crane operators, checkers/tallymen and lashers. Their main duties involve the following:

- Operators work on cranes (most often, there are two operators for an STS crane); in accordance with the Occupational Health and Safety regulations, operators work in two-hour intervals—e.g., an A operator 6.00 to 8.00 operating the crane while a B operator rests, and 8.00 to 10.00 the A operator goes to rest, and the operation of the crane is taken over by the B operator, etc.
- Checkers/tallymen are usually responsible for the following: checking containers according to the loading/unloading lists or tally sheets (numbers, positions), visual checking of technical status of containers (any damage must be reported), checking IMDG marking and seals, entering unloaded/loaded containers into the TOS system (through a mobile terminal); if any damage is noticed, a tallyman notifies (using the TOS system or by the radio) a foreman who supervises the loading/unloading process and/or a dispatcher who is responsible for insurance. OOG[1] containers (most often open top containers exceeding the parameters of a 40′ HC container) are handled according to a special procedure.
- Lashers (usually groups of two or three lashers) who perform tasks directly related to stowing (fixing on/taking off lashing and twist locks).

Figure 10.4 presents a scheme illustrating a process of loading/unloading a vessel by STS cranes, the stations taken by the particular employees and positions of the means of transport for containers (TT) in the operational area.

Considering the previously mentioned limitations, at the Luka Koper terminal, a Postpanamax container vessel (a mother vessel) is simultaneously handled by four

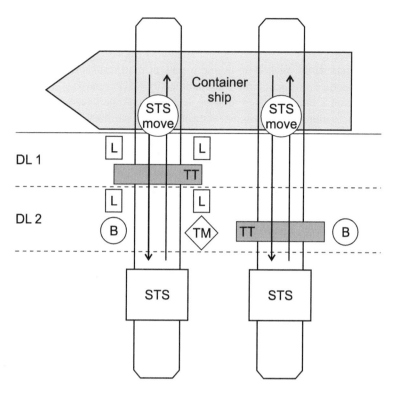

FIGURE 10.4 The scheme of a process of loading/unloading a vessel at the Luka Koper maritime container terminal.

Source: Based on Port of Luka Koper (2015, 24).

Legend:

STS—Ship to Shore
DL 1, 2—Drive Lane 1, 2
L—Lasher
B—Twist lock Basket
TM—Checker (tallyman)
TT—Terminal Tractor

STS cranes on average (by three cranes at the minimum and by five of them at the maximum). Smaller feeder vessels are usually handled by two gantry cranes on average. While handling 20′ containers and if possible, the cranes operate in a twin mode—that is, two containers are handled with one move of the crane. The cranes can also do double cycling, which can be problematic because then it is necessary to arrange import/export containers in a particular order. Double cycling is a combination of loading and unloading operations done during one working cycle of a gantry crane.

Loading and unloading operations take place according to an approved stowage plan, which defines the sequence of handling particular containers. Considering

optimization of processes, vessel stability and special stowage requirements applying to reefer containers, OOG and IMDG containers, the following assumptions have
been made:

- Containers are loaded into cells installed in the vessel holds:
    - Individual 20′ containers are loaded first in the single-lift cycle.
    - Next, all 20′ containers are loaded in the twin-lift cycle (heavy contain-
      ers are loaded first).
    - Then all 40′, 40′ HC containers are loaded (only to the slots specified in
      the stowage plan).
    - Containers stuffed with dangerous cargo, in accordance with IMDG
      requirements (only to the slots specified in the stowage plan).
    - Tank containers (only to the slots specified in the stowage plan).
    - Reefer containers (only to the slots specified in the stowage plan, with the
      possibility to plug them to the vessel power system).
- After that, the following containers are loaded onto the board of the
  vessel:
    - Individual 20′ containers and IMDG containers in the single-lift cycle
      are loaded first.
    - Next, all 20′ containers are loaded in the twin-lift cycle, then 40′ containers—
      in accordance with principle "from the seaside to the landside."
    - Heavy containers.
    - 45′ containers.
    - All OOG containers are loaded last.

One of the key aspects in the conventional model of terminal operation is the
proper allocation of storage yards for import and export containers that could
foster optimization of processes (shortening the time when vessels stay at the
terminal berths, increasing the speed of cargo handling processes, minimizing
redundant operations, shortening routes for terminal tractors). Hence, export containers are stored closest to the wharf (streamlining loading processes), whereas
import containers are stored closer to the railway sidetracks (to avoid conflicting
traffic of terminal tractors, RTG cranes and reach stackers during their simultaneous operations). Taking the optimization of processes into consideration, it is stra-
tegically advisable to arrange import and export containers in the way presented
in Figure 10.5.

Considering the change made to the concept of operational processes and their
functioning, a vertical approach (model) seems to be particularly interesting. In comparison to the traditional architecture of processes, the concept of applying vertical
drive in container handling operations at the BNCT assumes a total change to the
approach toward storing and moving containers around the yards. The significant
differences in the process architecture of the conventional (horizontal) and vertical
(BNCT) systems are presented in Figure 10.6.

During the process of loading, containers are delivered by means of land transport to the land side transfer area (LST), and then they are stacked in the automated

FIGURE 10.5   The arrangement of the storage yards for import and export containers at the Luka Koper container terminal.

Source: Reprinted with permission of Šik Sebastjan, Port of Luka Koper Head of PR Department.

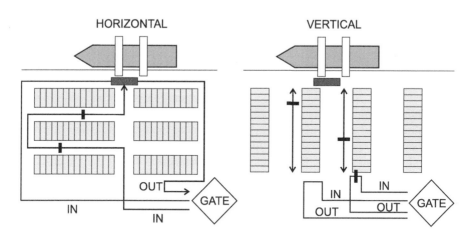

FIGURE 10.6   A comparison of the vertical model of process architecture at the BNCT (right) and the conventional system (left).

Source: Adapted and modified from Hutchison Ports Busan (2021).

stacking area (ASA) and moved to the water side transfer area (WST) with the use of gantry cranes (vertical movement). After that, they are transferred to the quay crane area (QC). During the process of unloading, the reverse sequence of operations is applied. The course of the processes is presented in Figure 10.7.

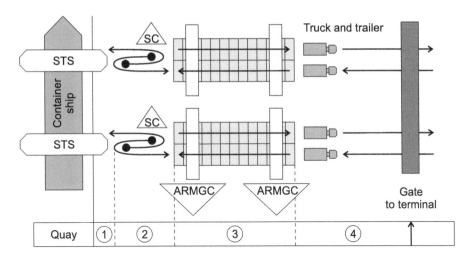

FIGURE 10.7   The course of container loading/unloading processes in the vertical system at the BNCT.

Source: Adapted and modified from Hutchison Ports Busan (2021).

Legend: 1. Quay Crane Area (QC), 2. Water Side Transfer Area (WST) with the buffer zone where straddle carriers (SC) operate, 3. Automated Stacking Area (ASA) equipped with automated rail mounted gantry cranes (ARMGC), 4. Land Side Transfer Area (LST).

The vertical approach allows operators to do the following:

- completely eliminate periods of waiting time in the QC, pickup and delivery areas (containers are not moved to other means of transport in the WST area)
- maximally shorten the time when terminal tractors with semitrailers stay in the area of the container terminal (the traffic is limited exclusively to the LST area)
- minimize handling operations performed for each container, owing to the fact that operations in the ASA area are automated
- eliminate problems related to terminal tractor drivers' poor navigation around the terminal yards (which is often a reason for congestions that cause collisions with the suprastructural equipment, such as STS cranes or reach stackers [RS])

As one of the first maritime container terminals, the CTA terminal in Hamburg underwent the processes of full automation at the beginning of the 2000s, which involved the use of fully automated gantry cranes and automated guided vehicles (AGV) and also one of the most advanced TOS-class systems. The requirements concerning low anthropopressure have prompted the CTA authorities to implement the concept of AGV electrification (which have been so far powered by diesel fuel), based on the FRESH[2] program. The authorities have also been prompted to design closer cooperation with low-emission modes of transport (such as railway—the Kombi-Transeuropa Hamburg Terminal) and to implement renewable energy sources (especially solar and wind power) in order to provide power to devices that are directly engaged in container handling processes. The electrification of AGVs has already resulted in lower emission and lower air pollution. The electrification of AGVs and the implementation of other ecological solutions have allowed the CTA to achieve high ecological efficiency, and as a result, it has been proclaimed to be the first maritime container terminal in the world that is neutral to climate (considering significantly lower emission of greenhouse gases, especially the reduction of CO_2 emission). This fact has also resulted in the necessity to implement changes to the standard TOS that would refer to the electrification issues.

Ro-ro terminals are characterized by some other specific features. They need to be equipped with ramps compatible with ro-ro vessel loading doors. Horizontal loading and unloading processes involving the rolling methods applied to trucks, semitrailers, roll trailers, cassettes or oversized cargo would not be possible without terminal ramps. Ro-ro terminals are also equipped with the infrastructure required for long-term storage of goods, with the possibility of changing means of transport, stuffing and stripping containers and other loading units.

During a typical process of shipping containers in maritime transport, handling a container from a vessel onto a semitrailer is done in the sequence of two maneuvers (see more in Gort [2009]). Logistics processes become more complicated when river barges are engaged—it is necessary to make two additional maneuvers. Transporting containers on a river barge means that there are additional operations related to intermodal processes and that some additional costs are to be generated (not mentioning the prolonged time of the entire operation). Therefore, it is important to find solutions aiming at the elimination of such inconveniences in operation of inland waterway ports, considering the fact that inland waterway and sea–inland waterway ports pursue higher operational efficiency. Hence, lower operating costs and efficient cargo handling between various modes of transport are highly significant. In order to achieve them, an IPSI (Improved Port-Ship Interface) project has been launched (IPSI Final Report 2008). The main aim of the IPSI project is to implement modern concepts pertaining to the development of intermodal terminals, vessels in short-sea-shipping (SSS) and also river barges in inland waterway shipping (Kaup and Chmielewska-Przybysz 2012, 501).

Considering intermodal terminals, the project is mainly focused on the development of automated cargo handling systems. Such systems should allow operators to efficiently handle the most common loading units in the world: standard ISO containers, container semitrailers, swap bodies and containers placed on special cassettes or pallets. Container handling at a maritime terminal is performed with the use of automated guided vehicles (AGV), which use cassettes especially designed for the IPSI system. The entire operation should be handled by an integrated IT (telematics) system.

According to the IPSI project, each IPSI terminal (depending on its hydrological conditions) should be equipped with one of two specially designed ramp systems (Kaup and Chmielewska-Przybysz 2012, 502):

- a permanent two-level ramp installed on the wharf to handle cargo during slight undulation of the water level/tides (up to 2m)
- a movable two-level ramp to handle cargo during high undulation of the water level/tides (up to 4.3m with the capability of tolerating higher levels)

The IPSI project assumes construction of two special types of vessels (see more in IPSI Final Report 2008):

- IPSI vessels[3] for short-sea-shipping
- IPSI barges[4] for river navigation in inland waterway transport

Despite the necessity of adjusting terminals and vessels/barges to the objectives of the IPSI project (which would obviously entail high financial investments), it is necessary to consider the fact that the previously mentioned solutions for IPSI terminals and vessels are relatively flexible. In fact, IPSI terminals can also handle standard ferries and ro-ro vessels (including Enisysvessels[5] or Interbarges).[6] IPSI vessels can call at any ports equipped with the basic port infrastructure and facilities for standard ro-ro cargo handling operations.

Hence, according to the IPSI concept, an IPSI terminal must meet the criteria of multimodality. It must be equipped with facilities to handle ro-ro processes and AGVs in order to achieve a standard level of container handling of 2 400 TEU per day or 400 TEU per hour at the reduced personnel costs (the permanent number of employees) (Lindstad and Uthaug 2003). It is assumed that containers handled at IPSI terminals will undergo automated (supported by adequate telematics systems) processes of identification, storage and handling, assuming the cooperation of three modes of transport (IPSI Final Report 2008):

- Inland waterway transport—river barges; however, only IPSI barges are taken into consideration because of their adjustment to the automatic systems of IPSI terminals (handling by AGVs).
- Railway transport—infrastructure and suprastructure to handle trains under the Rail Service Center (RSC).
- Road transport—infrastructure and suprastructure to handle tractors with semitrailers under the Truck Service Center (TSC).

An organizational scheme of an IPSI terminal with TSC and RSC is presented in Figure 10.8.

A terminal yard (a marshalling area) is divided into rows. On the left side, there is storage space for containers delivered by IPSI vessels and to be handled by the TCS and/or RSC. On the right side, there is storage space provided to containers delivered

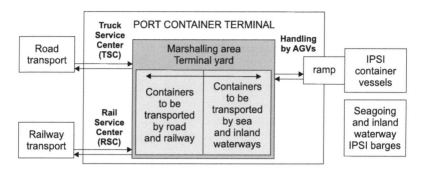

FIGURE 10.8   A scheme of an IPSI terminal.

Source: The authors' elaboration.

by IPSI barges. The AGV system should strictly cooperate with the system of auto-mated gantry cranes installed in the area of the terminal yard. On the one hand, the previously mentioned examples of applying various operational models in order to optimize operation of a container terminal indicate the use of innovative solutions that refer to the economies of scale or economic and managerial efficiency. On the other hand, they also implement factors that determine container handling operations performed at a terminal.

10.2   FACTORS DETERMINING TERMINAL OPERATIONS OF CONTAINER HANDLING

Permanently growing, the capacity of container vessels is clearly advantageous because it brings the economies of scale effects, generated by the marginal cost accounting and improved economics of transport processes. However, handling VLCS-/ULCV-type container vessels poses serious challenges related to efficient implementation of operations taking place at the terminal. Based on the previously mentioned operational models, the current pragmatics of the functioning of mari-time container terminals assumes the optimal use of wharves, sufficient numbers of necessary STS cranes, adequate capacity of storage yards for containers and the minimal loss of time during the operating cycles (e.g., gantry crane movement). Over the recent years, the external conditions have been changing dynamically in the field of operating bigger and bigger container vessels (more than 18 000 TEU). This fact means that it is necessary to provide adequate infrastructure and suprastructure at maritime port terminals. Larger and longer vessels may use entire cargo handling capabilities of a terminal. To maintain the assumed efficiency level that determines competitiveness, this disadvantageous phenomenon has to be com-pensated with adequate organizational changes in logistics processes and produc-tivity of logistics suprastructure of terminals (it especially refers to the operating cycles of terminal STS cranes). Generally, a typical operating cycle of an STS crane lasts for about 90 seconds, and in some particularly favorable conditions, it can be shortened to about 60 seconds (Yahalom and Guan 2021). However, in other cases when some hindering factors are involved (containers must be loaded into slots located deep down on the sea side of the vessel that is being handled), this time can be doubled. Considering the illustration presented previously, it is possible to analyze the results of changes related to the size of the vessels handled per operating cycles of the gantry cranes. A container vessel of the previous generation (6,000 TEU) is three times smaller than a Triple E vessel; hence, containers can be loaded into slots/bays in the following way: 17 containers in the width of the hold, eight containers in the depth of the hold (below the deck) and six containers above the deck; this means that there are 210 containers in a bay on average. The situation is different when the vessel capacity is 18,000 TEU. In such a case, a gantry crane must handle the following dimensions (considering the number of containers in a bay): the width is 22 (+5), the depth is 10 (+2), the height is 9 (+3) and, eventually, there are 376 containers in a bay, out of which only 210 containers do not generate any changes to the current parameters of the operating cycle of a gantry crane (no change). Still, the gantry crane must extend its trolley operation (further trolley) for 69 containers. It must reach deeper and higher (more hoist) for 76 containers, and 21 containers require more trolley and hoist operations (hoist + trolley). Such differences affect the average operating cycle of a gantry crane, and as a result, its efficiency/productivity is worse (crane productivity—P). It can be calculated in the following way (Moret and Lane 2016, 24):

$$P = \frac{L_K}{L_s \times l_c} \tag{10.1}$$

where

L_K—the number of containers;
L_s—the number of gantry cranes; and
l_c—the number of operating cycles per hour.

Modern gantry cranes can perform those additional movements in an average time of approximately two seconds (further trolley), and they need some seven seconds more for additional up/down movement. Hence, in the case of those particular containers, the operating cycle of a gantry crane is extended in time up to 92–98 seconds. After the aggregation of these numbers for the time required to handle the entire bay[7] (including 210 containers handled during the standard time of 90 seconds), the average time is 93 seconds (+3.3%). As a result, the number of potential[8] operating cycles performed by a gantry crane within an hour is reduced from 40 to 38.7 on average. These calculations are presented in Table 10.1.

Other (operational) solutions that significantly affect the productivity of gantry cranes are twin-mode (or triple-mode) and double-cycling operations. Twin-mode operations allow a gantry crane to handle two containers (2 TEU) simultaneously. At some terminals, gantry cranes perform triple-mode operations (and there are some plans to implement operations in which five containers are handled simultaneously).

TABLE 10.1

The Influence of the Size of Container Vessels on the Efficiency of Gantry Crane Operation at a Port Container Terminal

Change	Average growth of distance (m)	Time required (in seconds)	Container slots	Cycle time
No change	0	0	210	90
Further trolley	7.5	2	69	92
More hoist	8.25	7	76	97
Hoist + trolley	8.25	9	21	99
Average			376	92.3

Source: Data from Moret and Lane (2016, 26).

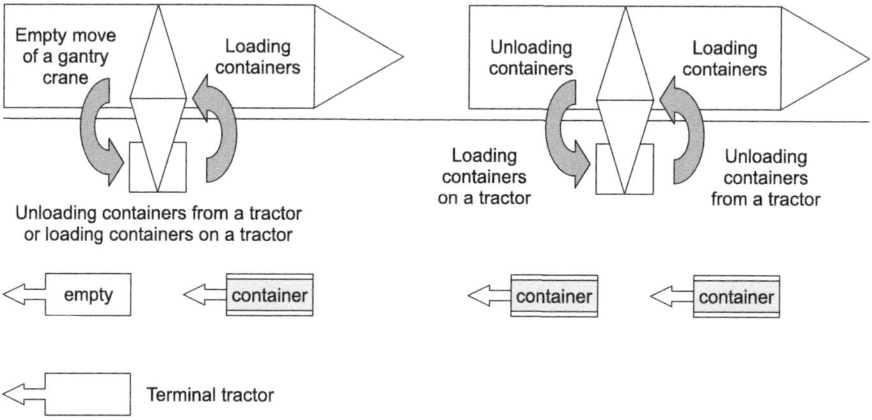

FIGURE 10.9 Comparison of Single- and Double-Cycling operations during the Processes of Container Handling Performed by STS Cranes.

Source: The authors' elaboration based on Zhang et al. (2015, 316–326).

This can radically increase productivity of gantry cranes and improve the total efficiency of a terminal.

Double-cycling operations are of a different character (although they also optimize productivity of gantry cranes). Double-cycling operations assume efficient movement of a gantry crane (with a container) in both cycles (shore to ship when loading vessels and ship to shore when unloading vessels). Such operations, by definition, are twice as efficient as standard single cycling operations. Additionally, such operations decrease the number of operations performed on the yard and the operational engagement of terminal tractors. A double-cycling operation is presented in Figure 10.9.

Considering further development and an increase in loading capacity of vessels (about 50,000 TEU), the previously mentioned assumptions cease to be rational. Therefore, today it would be advisable to look for new operational solutions that could be applied at container terminals and that would be adequate to the intended changes to the size of container vessels. One of the concepts considered in this context (which could significantly modify the current model of vessel handling processes at a container terminal) is the concept of handling a container vessel from its both sides simultaneously. It would eliminate the need to rotate the vessel during loading and unloading operations.

Another solution is a multistory container-stacking system, which is still being (conceptually) developed by DP World. It allows operators to keep containers on the storage yard in special, individual, separate rack compartments (for each container). The system doubles the capacity of the storage yard, and at the same time, the speed of operations is increased, as well as the availability of individual containers (there is no need to stack containers and move them in order to get a container located at the bottom of a particular stack), safety and efficiency of handling operations. The implementation of such a system would, however, entail a complete change to the current layout of terminals and reconfiguration of processes, along with changes to the suprastructure (new types of container handling facilities).

Presented previously, the examples and concepts of optimization of processes related to container vessel handling operations taking place at terminals clearly indicate the significant role of terminal infrastructure and suprastructure.

10.3   TERMINAL INFRASTRUCTURE AND SUPRASTRUCTURE

A port container terminal comes as a set of infrastructural and suprastructural elements that offer the synergy effect, leading to efficient operations of handling cargo and means of transport. The fundamental tasks of terminals involve the implementation of various loading and unloading operations for means of transport taking part in numerous transport relations, short-term storage of containers at storage yards and the performance of other operations involving containers and cargo if necessary (e.g., consolidation of general cargo into the form of containerized cargo, LCL-FCL, or the other way around). The complexity level of various transport operations results in the necessity of applying special devices and providing access to transport infrastructure (internal and public roads, railway tracks, port basins and wharves).

Each port container terminal uses port infrastructure, which is the material basis that determines the level of port service production (Grzelakowski and Matczak 2013, 36; Misztal, ed. 2010, 33). A terminal is composed of the water area and the land area. The water port infrastructure includes a roadstead, an outport, port canals and port basins. A roadstead is an area where vessels wait for the permission of the port authorities to enter the port. The outport is the first part of the port where vessels can maneuver and move along port canals to get to a designated port basin and a particular berth, where cargo handling operations are going to take place. The parameters of the water port infrastructure determine the capability of a container terminal to handle vessels of particular sizes. The basic technical and operational parameters are as follows:

- The maximal depth of the roadstead, outport, port canals and port basins— it determines the size of vessels that are going to be handled at the port (deeper elements of the infrastructure allow vessels of larger draught and linear dimensions to enter the terminal).
- Bigger dimensions of the outport (length and width), port canals and port basins allow operators to handle big vessels (at present, middle-size and large container vessels are operated more and more often).

Over recent years, an increase in the shipping capabilities of container vessels has been observed. It results from the fact that shipowners have been implementing the economies of scale into their production to increase opportunities of shipping service production (Szwankowski 2000, 29). Pursuing such a developmental strategy indicates the necessity of operating bigger and bigger vessels that can be handled at container terminals located at large maritime ports with adequately developed infrastructure. Keeping up with fast development of maritime container shipping requires port authorities to make investment decisions that could improve accessibility of their ports to large container vessels. However, the process of investing into water infrastructure involves considerable financial expenses and time. Furthermore, it also depends on elements of land infrastructure, which sometimes needs modification as well.

The land infrastructure of port container terminals includes an area which is developed in a way that allows operators to efficiently change means of transport and to perform indispensable terminal operations. The land infrastructure of a container terminal includes the following:

- wharves
- terminal transport roads
- terminal railway tracks
- power, telecommunication and water and sewage systems
- IT systems (to provide access to the intranet and the internet)
- supervision and security systems (terminal monitoring systems)

The basic element of a terminal is its wharf, forming the shoreline with its quays, piers or port canals with the adjacent port areas.[9] The wharf of a container terminal is defined by its length and width. The width of terminal wharves can be of quite a considerable size, considering the necessity of locating vast storage yards for containers (sometimes the width may exceed several hundred meters). A parameter that determines capabilities of a terminal to handle vessels is the depth of water at the terminal wharf as it defines the size of vessels that can be handled at the particular berths (the deeper the water, the larger vessels can be handled there). Mooring and fender/bumper facilities come as supplementary equipment at the wharves. They are used for immobilizing vessels that stay at the port during cargo handling operations and also for providing safety to vessels (vessels may drift with waves, gusts of wind or waves generated by other vessels passing by).

The efficient operation of a terminal requires proper maintenance of technological transport, which ensures the flow of containers between various means of transport, cargo handling facilities, storage yards and warehouses. Hence, it is necessary to design a network of internal roads that will be used for transporting cargo. Internal roads should be communicated with the external (public) infrastructure to receive and to send containers by road transport. The course of the internal road infrastructure of a terminal should be optimized adequately to its size and methods accepted to handle cargo and means of transport. Terminals may be also equipped with railway tracks that are used for receiving and sending containers in various transport relations.

The functioning of a container terminal also depends on power, telecommunication and water and sewage systems. Most static cargo handling facilities are powered by electricity and require numerous electrical connections. Access to electrical power is also required for all-purpose and special-purpose warehouses, stations for reefer containers that must provide and maintain conditions specified for storage of particular cargo in external conditions. The connection to the water and sewage system comes as an obvious condition, considering the necessity of installing firefighting systems, rainwater drainage, maintenance of proper standards at office and social buildings for employees. At present, telecommunication cable networks are standard equipment that must be provided to maintain internal and external communication at a terminal (it is possible to distinguish internal telecommunication and public telecommunication networks).

An IT system comes as a critical part of the infrastructure because it is responsible for all cargo handling operations implemented at a terminal. Any failures of IT systems paralyze all other cargo handling and maneuvering activities because it is not possible to identify particular containers (what cargo is stuffed inside). Therefore, it is significant to design, implement and maintain IT systems of a terminal in the proper way. The IT infrastructure can be based on multilevel systemic solutions that allow operators to manage various functional areas of the terminal. Each terminal must have an operating system to supervise the course of all operations taking place at the terminal: the terminal operating system (TOS).

Considering the specificity of port cargo handling facilities, all terminals are treated as state borders, and customs clearance operations take place within their premises. Hence, it is necessary to monitor the area of a terminal by terminal security services and monitoring systems. Seaports and all the objects within their premises are considered as areas that require protection (*Journal of Laws* 2008; EC Regulation no. 725/2004 2004). The authorities of terminals are obliged to provide security to their employees, vessel crews, truck drivers, freight train crews and office employees. Apart from protection provided to people who are present in the area of a terminal, it is also necessary to provide protection to technical equipment and cargo against any interference of unauthorized parties.

The suprastructure of container terminals comes as technical equipment indispensable to run business operations (Misztal, Kuźma and Szwankowski 1994, 49). Terminals operate efficiently if proper infrastructure is provided and supplemented with adequate suprastructure. The following suprastructural elements can be listed at container terminals:

- cargo handling facilities
- storage yards
- warehouses
- terminal (office and technical) buildings

Cargo handling facilities determine the efficiency of cargo handling operations. Therefore, each container terminal pursues its optimal investment policy to acquire facilities that are needed. Handling containers requires specialized equipment, which is technically adjusted to handle and to transport containers. Frequently, the authorities of a terminal also decide to purchase supplementary all-purpose facilities that can be used for handling heavy and oversized cargo.

Typical cargo handling facilities are as follows:

- shore-to-ship cranes (STS)
- rail-mounted bridge cranes (RMG)
- rubber-tired gantry cranes (RTG)
- container straddle carriers (CSC)
- reach stackers (RS)
- self-propelled forklift trucks for containers
- self-propelled forklift trucks for pallets
- terminal tractors (TT)
- terminal and road semitrailers
- roll trailers
- self-propelled all-purpose cranes

The characteristic elements of each port container terminal are shore-to-ship cranes used for fast vessel handling. Usually, one STS crane is treated as one cargo handling station, and depending on the wharf length, there might be several or even more than ten such stations located on the wharf. The size of STS cranes depends on the size of vessels that are handled at a particular terminal. The operating range of the boom of an STS crane is a determining factor here as it allows the crane to reach containers placed in the longest distance from the shore line. The lifting capacity of STS cranes is similar and adjusted to the standards specified for container weight, enlarged by a safety margin (60 tonnes approximately). Maneuvering operations on storage yards are usually performed by self-propelled vehicles that move and deliver containers to the places specified by the TOS. The largest group is constituted by vehicles for transporting and stacking containers; however, during warehouse operations involving general cargo forklift, trucks are most commonly used. Containers are taken from the STS cranes to their destination storage areas by terminal tractors that are usually equipped with semitrailers or roll trailers. The use of particular cargo handling facilities during cargo handling operations is presented in Figure 10.10.

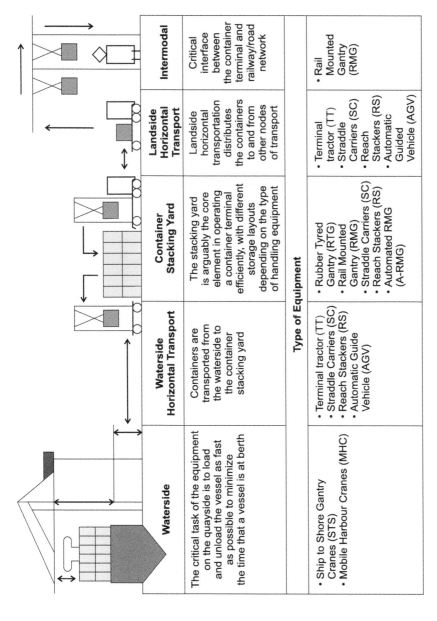

FIGURE 10.10   The use of cargo handling facilities in the horizontal structure of processes at a maritime container terminal.

Source: The authors' elaboration based on ALG (2001).

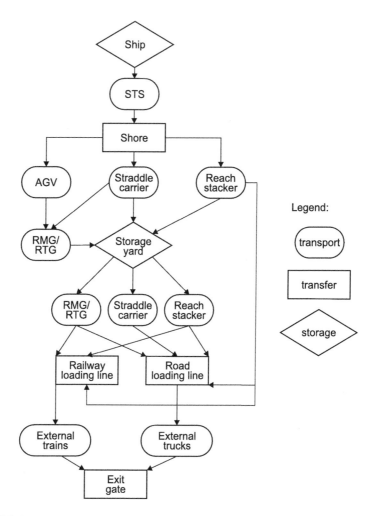

FIGURE 10.11 A process map presenting the use of cargo handling suprastructure at a maritime container terminal.

Source: The authors' elaboration based on Kubowicz, D. (2019, 487).

The use of cargo handling facilities at a maritime container terminal in the form of a process map is presented in Figure 10.11.

All terminals have storage yards where containers are kept, waiting for transportation, customs clearance, sanitary inspections or any other procedures. Storage yards are a part of terminal suprastructure, and they have to be prepared in a proper way. The surface of a yard where containers are stored must be reinforced to withstand the pressure of container stacks and maneuvering operations of transport and cargo handling vehicles. Cargo and container handling operations also take place at night; hence, it is necessary to provide adequate lighting. Some containers need special treatment, and they must be stored at special areas designated for keeping containers with hazardous cargo or reefer containers. Reefer containers must be kept at the stations equipped with electrical power connections and generators. Depending on the size of a terminal, there might be one or several storage yards that sometimes are used for other various purposes.

At terminal premises, there are also all-purpose warehouses or freight shelters for general cargo—non-containerized goods. Such objects can be observed at terminals that were built a few decades ago to consolidate general cargo into containers directly before loading them onto a vessel. Modern container terminals do not have all-purpose warehouses because such objects take space that could be used for storing containers. All-purpose warehouses can be still useful if a terminal handles bulk or semi-bulk general cargo or heavy cargo (such as formed steel components or paper bales). Apart from warehouses at a container terminal, there might be office and technical buildings located as well. Office buildings obviously accommodate offices and rooms for computer servers of the terminal IT systems that supervise all the cargo handling and auxiliary operations.

Proper infrastructure and adequate suprastructural facilities allow maritime container terminals to provide a wider range of logistics services.

10.4   SERVICES PROVIDED BY PORT CONTAINER TERMINALS

10.4.1   STUFFING AND STRIPPING CONTAINERS IN MARITIME TRADE

In maritime trade, the process of stuffing containers involves filling them with cargo that is to be transported by sea. Stripping containers means that they are emptied and the goods transported inside are simply taken out. Stuffing containers with cargo may take place at the premises of manufacturers/shippers of the goods that are going to be transported in containers. The process may also take place at a port container terminal. Both processes, stuffing and stripping containers, require expensive equipment and reinforced surface that can withstand the pressure up to 25 tonnes per m^2. If the container turnover is low, such expenses are pointless. It is more economically justified to deliver goods by tarp-covered trucks or by railway cars to port warehouses and to stuff them into containers there. It is particularly advisable when the cargo of one supplier does not fill the entire container. The situation is similar when containers are stripped. If the stripping takes place at the port premises and the goods are not delivered to their consignee in containers, then costs related to transporting empty containers back to the port can be avoided. Such situations can be often observed in Poland and other countries (Dąbrowski, Kaliszewski and Klimek 2013). Manufacturing factories that send their products in large quantities in containers—for example, automotive components—can stuff containers at their own premises. Similarly, consignees of large quantities of products delivered in containers, such as car assembly plants that receive automotive components, can strip containers at their premises as well.

Regardless of a place where those operations are performed, the core of the container stuffing process is the optimal preparation of the cargo for being transported in a container. It involves the following activities (Wiśnicki, ed. 2006, 115):

- stuffing the cargo into a container
- stowing the cargo in a container
- lashing the cargo in a container

When containers are stripped, the lashing elements are taken off and containers are stripped.

Proper stuffing is important not only for the safe transportation of the cargo inside containers but also for safety of people who are involved in the operations performed on particular containers—for example, handling a container might result in a damage done to the means of transport or container handling facilities at a terminal. Container stuffing operations depend, first of all, on the characteristics of the cargo, types of unit packaging, stowing and lashing technologies applied to secure cargo inside containers. Goods requiring ventilation must be transported in ventilated containers and respectively: goods requiring cooling, in reefer containers; bulk dry cargo, in containers dedicated to dry bulk cargo; heavy cargo, on platform containers; liquid cargo, in tank containers; etc. Most often, goods stuffed in containers are packed into cardboard boxes, cases for heavy cargo, pallets, sacks or barrels. The size and durability of the packaging determine the number of pieces that can be stuffed into one container—for example, one 40′ container will contain 8,000 pairs of shoes in cardboard boxes or 25 evenly distributed Euro-pallets and 21 industrial pallets. It should however be remembered that the maximal acceptable gross weight of containers (e.g., typical 40′ containers) is 26 tonnes.

Goods inside a container must be properly stowed and any free space between the particular pieces should be filled with dunnage materials, such as wooden planks, beams, foam, Styrofoam, air bags, cardboard, waste tires, etc. The floor of a container is equipped with lashing rings to stow the cargo to prevent its movements inside the container.

All stowage operations should be started with a preparation of a loading list, which is a list of all the goods commissioned by their owner or by a forwarder for transportation in containers. A loading list is usually based on a booking list, which, in turn, is a basis for developing a stowage plan for an individual container. A stowage plan indicates where and how to load the cargo into a particular container.

While developing a stowage plan, a few fundamental stowing principles must be taken into consideration (Wiśnicki, ed. 2006, 145):

- The mass center of the stuffed container should be close to its geometric center.
- The cargo should be evenly distributed on the container floor.
- The height and the way of stacking the cargo should be specified.
- The mechanization level of cargo handling operations should be specified—namely, it should be indicated whether and what type of mechanical devices will be used during stowage operations.
- It is necessary to provide the safe and easy stripping of the container at its destination place.

The goods inside a container are rigidly and firmly stowed in order to prevent any movement, tilting or falling. There are two types of rigid stowage (Wiśnicki, ed. 2006, 172):

- blocking stowage—that is, rigid stowage immobilizing the cargo units packed next to each other, with the use of stowing dunnage but without lashing fittings
- individual stowage that is applied to provide rigid fixing of individual packaging units with the use of lashing fittings, such as ropes, chains, textile straps, metal tapes, etc.

Acting in accordance with the previously mentioned rules depends on properly trained and experienced stowing teams. The methods of stowage mechanization range from manual packing, mechanic packing (with the use of fork lifts) to automatic packing (with the use of automatic devices). An example of an automatic system for loading pallets into a container (or generally, a semitrailer) is SkateLoader (Europa Systems 2021). This is a permanent system for loading pallets into a container that is installed at a particular place of loading/unloading platform or abeam and which can handle several docks. Mounted on a steel ramp, transporting platforms lift and move the cargo. The main function is automatic loading and unloading of the entire cargo at one go (during one process). The idea of how that system operates is presented in Figure 10.12.

Usually, containers intended for stuffing operations are provided by shipowners together with container seals. After the container has been stuffed with the cargo and its door has been closed, the seal is placed on special locking rods fixed to the door

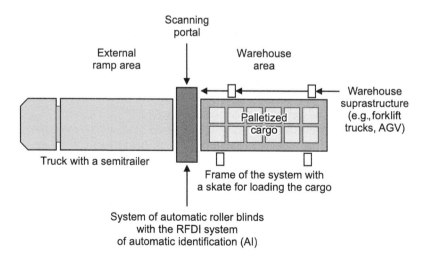

FIGURE 10.12 The SkateLoader automatic system for loading containers.

Source: The authors' elaboration based on Europa Systems (2021).

panels.[10] Customs (customs and tax) authorities and institutions responsible for food and veterinary inspections may put their own seals on containers (customs, phytosanitary seals, etc.). Apart from stowing the cargo inside containers, it is also necessary to consider their safe transportation by sea, so the problem of stowing containers on a vessel becomes particularly important.

10.4.2 STOWING CONTAINERS ON BOARD

After they are loaded onto a vessel for transportation, containers must be securely placed and stowed. The proper stowage of containers on a vessel must follow the rules described in the previous part of this chapter (a conventional operational model), but also, some important elements of stability safety must be taken into consideration during that process. Considering hydro-meteorological conditions (wind, waves), cargo transported by sea may be affected by various forces that can cause its movement and a change to the center of gravity, which can, in turn, affect the stability and the metacentric height of the vessel.

Hence, container stowage is extremely significant not only to the safety of the cargo inside containers but also to the safety of the vessel itself. According to the current regulations, containers that have not been stowed should never be allowed on board. During the cargo stowage operations, the sequence of cargo unloading from bays should be taken into consideration. Nevertheless, the stability of a vessel is always more important than the optimization of cargo handling operations. Hence, failing to meet the requirements of proper stowage (both the cargo inside containers and the containers on a container vessel) may be a cause of serious problems during the sailing, and it can result in destruction of the cargo and/or loss (sinking) of containers or even in the sinking of the vessel in some extreme cases.

Considering the stability requirements, the following rules should be applied (Kuhne+Nagel 2021):

- The heaviest containers should be placed at the bottom of the hold of a vessel whenever possible.
- While containers are being loaded, metal guide rails should be used whenever possible.
- Considering the vertical section, heavier containers should be placed in the bottom layers and lighter ones in the upper layers.
- Considering the horizontal section, light containers should be placed along the sides of the vessel, and heavy containers should be placed near the center of gravity of the vessel;
- The door panels of the containers loaded on the vessel should be oriented toward the stern.
- If there are two 20′ containers loaded on the spot dedicated to one 40′ container, their door panels should face each other.

Loading containers takes place according to a stowage plan that is developed individually for each container vessel waiting for the cargo at a terminal. The stowage plan indicates places dedicated to the particular containers on the vessel, namely a particular stack, layer and row for each individual container. Each container position in the loading space of a vessel is described with a six-digit code. The first two numbers are defined as BN (Bay Number), and they refer to the number of the cross section of the vessel, from the bow to the stern. Containers that are 20′ are numbered with odd numbers, and 40′ containers that are placed on two smaller containers are numbered with the arithmetic mean of those loading units. The next two digits of the code define the position of the container in the beam of the vessel, and they refer to the particular rows (RN—Row Number). The containers placed exactly on the axis of the vessel are numbered with 00. The containers placed of the left side are marked with even numbers, whereas those placed on the right side are marked with the odd numbers. The last two digits of the code define the number of layers. Each layer is described with the subsequent even number. The numbers of rows and layers under the deck and above the deck depend on the size of a particular container vessel (Bartosiewicz 2020, 77–80).

Despite the fact that there are various types of container vessels (as described in Chapter 2), the construction and dimensions of corner castings required to lift, fix and lock containers in stacks have been standardized. Owing to the construction and durability of the container frames, the weight of stacked containers is transferred to the corner posts and not to the container tops. Generally, it is possible to stack loaded containers into nine layers and empty containers into 12 layers. The integral parts of a container frame are four bottom and four top corner castings, which allow containers to be locked in a stack with a simple device referred to as a twist lock. The bottom layers on the board of a container vessel are connected with the use of a lashing turnbuckle system.

The optimization of container handling operations taking place at maritime container terminals requires synchronization of processes that are managed by various entities (stakeholders of the process of container transportation). Most often, the basic (indispensable) synchronization level is achieved by providing access to a terminal operating system to consignees of containerized cargo.

10.4.3 Access to the Operational System of a Terminal for Cargo Consignees

In order to achieve the optimal level of synchronization in their operations, operators of container terminals who use integrated operating systems (TOS) provide free but limited access to some functions (only indispensable ones) to external entities (including cargo consignees) through an access platform, under the regime of the full identification of the user.

Considering the complexity of processes taking place at a port and in the designated terminal structure, the synchronization range of the TOS is varied; however, it can generally refer to the integration with other systems of the following types:

- WMS (Warehouse Management System), when storage processes are considered
- TMS (Transport Management System), including vessel systems (and stowage plans) and notifications of trucks/trains, to improve the arrival and

departure of means of transport to and from the terminal (with the function of Optical Character Recognition—OCR)

- MES (Manufacturing Execution System), when logistics operations taking place at a port under logistic supply chains are considered (with the option of integration with the ERP—Enterprise Resource Planning)
- YMS (Yard Management System) to streamline terminal (operational) processes
- AMS (Asset Management System) to improve the use of suprastructure vehicles
- RMA (Return Merchandise Authorization) when handling reverse logistics is taken into consideration
- B2B (Business to Business) platform and access to EDI (Electronic Data Interchange)

The Synaptic TOS (Synaptic n.d.) is an example of the TOS-class software that provides access to the previously mentioned functions. The Synaptic TOS is based on 3D technology that uses the Microsoft SQL database server and its own application server.[11] The system cooperates with modern platforms of operating systems, such as Android, Windows or iOS. The YMS version of the system is optimized to cooperate with a wide range of Internet browsers, including Google Chrome, Mozilla Firefox, Microsoft Edge, Internet Explorer and Safari (both in the desktop and mobile formulas). The system can be integrated with other systems provided by software developers, so automatic information flow and optimization in the entire supply chain are provided. Mobile, portable hardware devices (portable terminals) can be installed in cockpits of RSs, STSs, RTGs, TTs and of other vehicles maneuvering on terminal yards. The optimization of management of operations taking place on terminal yards is implemented through analytical modules and operation monitoring (deployment of containers and suprastructure vehicles) and integration with external ERP-class systems or devices, such as automatic scales and gates (the use of ORC).

In general, the features of functionalities and accessibility of the terminal management software for external entities (with the particular consideration of cargo consignees) should include the following (Synaptic n.d.):

- managing cargo handling processes and internal processes of a terminal
- providing advanced algorithms for controlling distribution of containers on terminal yards
- automation of maneuvering processes based on receiving and releasing containers
- an interactive 3D map
- an algorithm for calculating maneuvering operations
- digitalization of documents, customs records and other records—for example, technical, phytosanitary or veterinary records
- automatic register of services referring to containers—for example, repairs, washing, painting, etc., and PTI inspections
- an internet application for B2B customers
- the possibility of receiving a quality control order

- integration under the EDI standard
- the possibility of customization and implementation of dedicated reports

Such a wide offer of operations related to container handling at maritime container terminals is also related to the fact that numerous fee elements have appeared, which reflect the scope of the use of available/offered services.

NOTES

1. Out of Gauge (OOG) is a common term to describe oversized cargo that cannot be transported in a standard 40′ HC container (High Cube of extended height).
2. The objective of the FRESH program is to provide compatibility of electrified AGVs with the German power system in order to achieve amperage standardization (50Hz) and to achieve additional capacity to store electric power during the time of lower demand.
3. An IPSI vessel is a one-hull vessel equipped with ro-ro stern ramps along the full beam of the stern. It can be also equipped with two or three decks of straight load lines and internal ramps that allow operators to move cargo quickly and efficiently. There are also devices for automatic stowage note.
4. An IPSI barge is equipped with a ro-ro bow door, one or two decks of straight load lines and devices for efficient cargo stowage.
5. An Enisys vessel—the construction of such vessels allows operators to reach high speed at low fuel consumption. Equipped with the Azimuth technology, such vessels meet multimodal and intermodal standards (involving containers, trailers, swap bodies, cassettes), including automated stowage. Owing to these features, Enisys vessels are also very efficient in navigation and port maneuvering. They are also ecologically friendly, low-emission vessels.
6. An interbarge is a river barge dedicated to efficient ro-ro technology handling. Owing to the new technology and hull design (lower demand for steel, lighter construction, higher displacement and small draught), interbarges are characterized by high loading capacity, efficiency and simplicity of use during ro-ro operations.
7. A bay is a block of containers stacked on one another. In the IT systems applied to handle vessels, containers loaded in this way are referred to as bays. Hence, a stowage plan of a container vessel is made of the particular bays.
8. The ultimate results depend on the operator of a gantry crane and their skills; generally, the influence of the human factor on the discussed parameter causes a decrease in the potential productivity calculated in the way presented above even by 20%–30%.
9. A pier is a hydro-technical object where warehouses, cargo handling facilities, transport roads, railway tracks and other components can be located. Usually, piers and quays are light constructions, on which small storage facilities can be based, along with access roads. The banks of port canals can be used for mooring and handling vessels—as it can be observed at ports located at river estuaries. See more in Misztal, Kuźma and Szwankowski (1994, 45).
10. The seal may be removed exclusively with the use of metal-cutting shears, and it forms the first line of protection against an unauthorized opening of a container door during transportation.
11. The system is made of a desktop application for PCs, laptops and tablets and a mobile application for warehouse terminals, smartphones and data readers. It is also possible to connect the system with the internet application, Synaptic.Web, dedicated to mobile devices through a dedicated SIM card that provides an access to the GSM system in the real time.

11 Information and IT (Telematics) Solutions in the Management of Containerization Processes in Maritime Transport

11.1 TASKS AND ARCHITECTURE OF TELEMATICS SYSTEMS IN THE MANAGEMENT OF CONTAINERIZATION PROCESSES

At present, containerization has already become a domain of broadly understood global shipping processes that are supported by information technologies. Hence, considering its role in global trade, maritime transport—largely represented by container unit shipping—requires dedicated IT support systems (because of the immanent features of this mode of transport).

Considering huge volumes of containerized cargo transported by sea, the maintenance of the acceptable levels defining operationalization, synchronization and optimization of processes based on containerization is inextricably related to the dynamic development of potential and dissemination of IT/ICT technologies. IT/ICT technologies are based on advanced telematics systems. Telematics comes as a notion from the field of information technology and, considering its present state-of-art level, it is applied mainly in relation to (see more in Wydro [2005, 116–118]; Miler [2019, 157–158]):

- structural solutions in which digital communication and digital acquisition and processing of information come as the integral elements of the system, which are constructed in accordance to the requirements of that system
- technical solutions in which universal telecommunication and IT systems are applied in an integrating way

Therefore, for the requirements of this monograph, it can be assumed that telematics refers to largely integrated telecommunication, IT and automatic control solutions adjusted to the needs of physical systems, resulting from their tasks, infrastructure, organization, maintenance and management processes.

DOI: 10.1201/9781003330127-15

As the previously mentioned information clearly indicates, the most important functions of IT systems are those related to information operation. This refers to acquiring, processing, distributing, transmitting and using information in various decision-making processes. These include processes that are implemented in a determined way (e.g., automatic control) and processes that result from ad hoc and discrete situations (decisions made by consigners, dispatchers, operators and independent users of a particular infrastructure). Telematics systems and applications are developed for specific processes (e.g., to transport containerized cargo by sea). Another important feature of telematics applications is their ability of the efficient associating of operations performed by various subsystems and putting them into a coordinated (integrated) mode of operation, which significantly affects another feature of these systems—their scalability (Miler 2019, 159–160).

Providing information in telematics applications is implemented in an automatic way (usually under the process of constant monitoring) or in an interactive way upon the user's request. Monitoring software is most frequently applied to collect information about the quantity and quality of shipping processes (including information related to containerization processes), with the consideration of using such information to make decisions involving a choice of optimal economic solutions or actions adequate to situations where there is too much risk to human life and property or natural environment.

Therefore, considering such an approach, maritime containerized transport must undergo analytical procedures that can increase the potential of optimization in shipping processes and auxiliary activities (cargo handling operations, storage, consolidation, etc.). The analytical and decision-making processes are not possible without adequate systems for acquiring information referring to navigation, port operations, environment, such as generating, gathering and transferring information (see more in Bujak, Smolarek and Orzeł [2013, 216]), analytical and decision-making elements (processing information with the use of smart, advanced decision-making algorithms) and executive elements (transferring managerial information—decisions to be implemented by the subordinate units of the system). Understood in such a way, a telematics system (a monitoring system of functional elements of smart maritime transport—for example, a system of container handling) should be composed of at least three modules (see more in: Miler 2019, 160; Miler 2011, 210):

- a sensory module—to acquire information from available sources (e.g., automated identification systems [AIS], long-range identification and tracking systems [LRIT], Safe Sea Net that monitor means of transport and location of containers with the use of the Global Positioning System [GPS] and movement or humidity sensors and other similar sensors that monitor technological regimes of container shipping, including smart camera and computer vision systems or digital twin technology solutions)
- an analytical and decision-making module—to eliminate information entropy and to process information into decisions (based on smart decision-making algorithms, such as the Critical Path Method [CPM], Program Evaluation and Review Technique [PERT], payoff matrices, the game theory, data mining, smart agents, machine learning, elements of artificial intelligence, including others, and so called triggers that are installed in the software supporting the analytical process)

- an effector module—to transfer decisions (most preferably in an automated way) for immediate execution (with the consideration of preventive functions referring to the categories of safety and to the optimization of operational actions)

The results of the implementation of telematics systems in maritime transport (in relation to the container handling systems) are, first of all, of qualitative nature, and considering a holistic approach, they contribute to the following (see more in: Miler 2016a, 135–136; Miler 2016b, 385–386):

- improvement of safety and security of maritime transport and container turnover (in the domains of safety and security—through the complete identification of containers, their cargo, beneficiaries/consignees/consigners in the entire logistic-transport chain)
- improvement of parameters defining the economics of maritime transport through the monitoring of container handling and shipping processes, with the identification of anomalies in the process that has been planned (its duration period, route, technological regime, shipping parameters, such as, for example, humidity, temperature, unauthorized opening of the container doors, sudden delay after a dropping of a container) which can allow insuring parties to identify precisely the liable party to claim compensation (it generally results in lower insurance premiums)
- improvement in the operational efficiency/productivity of maritime port terminals, with the particular consideration of container terminals, through information cooperation of TMS/FMS (Transport Management/Fleet Management) systems, including cargo applications (such as vessel planning) with TOS-class (Terminal Operation) systems;
- improvement in the availability of maritime transport and a higher share of this mode of transport in logistic supply chains through specialized handling of containerized cargo (with the consideration of stuffing, consolidation, deconsolidation and stripping processes pertaining to containerized loading units)
- higher integration of maritime transport in intermodal and multimodal shipping, with the particular use of inland waterway transport in the process of integrated container shipping by inland waterways, involving the IPSI (Improved Port/Ship Interface) technology

To sum up, telematics systems are applied in maritime containerized transport to acquire, to process, to present and to transfer information (more and more frequently, the access to the required information in real time becomes a standard, e.g., data provided in the cloud, Internet of Things, block-chain technologies) in four fundamental fields of functionality, understood as follows:

- monitoring systems of container loading units
- loading systems in maritime containerized transport (of the vessel-planning class)
- terminal operation systems (TOS)
- access and integration systems

11.2   MONITORING SYSTEMS OF CONTAINER LOADING UNITS IN MARITIME TRANSPORT

Considered in the global scale, international trade takes place every day, with the use of millions of containers, regardless of the season, weather conditions, means of transport or stages of logistic processes. Considering the economic point of view, any mistake made in the proper (adequate to a particular, defined path) container handling triggers direct consequences, such as increased operating and transactional costs, elongated operations, increased risk of further mistakes. It can also have indirect consequences, such as untimely delivery of containers and their cargo, problems with the continuity of processes, a limited commercial offer, lowered commercial reliability, loss of reputation on the market or a lower level of competitiveness.

Unfortunately, while shipping millions of containers in global processes of maritime transport, it is possible to observe a needle-in-a-haystack syndrome too often. Too few containers are inspected and too few transport and forwarding companies apply advanced container monitoring systems that not only do not hinder logistic processes (as long and time-consuming inspections do), but they streamline and accelerate container handling processes.

Tracking containers, monitoring all specified anomalies (such as the opening of container doors, damages to container wall panels, changes to temperature, etc.) in real time, theft detection and reporting the transit status (a stage in the logistic chain) are only some of potential capabilities provided by modern systems of container monitoring that make them some of the most important tools for management of logistic processes in maritime transport.

Modern systems have to meet the requirements of container monitoring in real time and also the requirements of immediate accessibility of tracking/monitoring data provided to all the stakeholders of the system, starting from entities of maritime transport, maritime operators and forwarders and ending with relevant inspecting agencies of the countries involved in the process. Therefore, integrated systemic solutions applied in the discussed field have to be provided with the following tools (see more in: Miler 2016a, 135–136; Miler 2016b, 385–386):

- optimization tools—to improve the efficiency of logistic processes and to lower their costs
- technological tools—to provide the possibility of monitoring individual containers in the entire logistic process through the use of localization technologies
- sensory tools—to provide monitoring and alarming about anomalies/deviations from the parameters specified for the container handling processes (e.g., the route, technological regimes defined for shipping, such as temperature, humidity, exposition to light, etc.) through the installation of individual sensors or multi-sensors
- integration tools—to use current positioning technologies (GPS) and automated identification (AI) bar codes (BC) and radio frequency identification (RFID) technologies

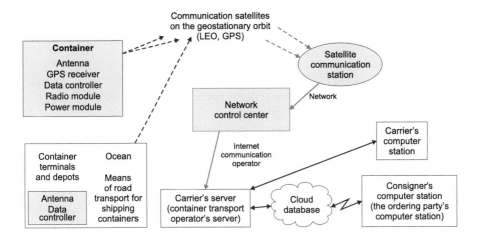

FIGURE 11.1 A functional scheme of the Electronic Container Tracking Service system.

Source: Authors' elaboration adapted and modified from Ahn (2005, 1722).

Modern systems of container monitoring are referred to as ECTS (Electronic Container Tracking Service) systems, and they use satellite LEO (Low Earth Orbital) systems under the GPS positioning. The information can be transmitted with the use of an adequate interface compatible with the IT systems of shipping companies and maritime transport operators (and even transmitted to the systems installed on mobile devices). A scheme of the process of container monitoring and identification through the ECTS (Electronic Container Tracking Service) system is presented in Figure 11.1.

Electronic security for the integrity of container door closure and container monitoring systems more and more often rely on the RFID technology. The main task of ECTS-class systems understood in such a way is to provide security to container shipping in the entire logistic supply chain. The managing party must be provided with the access to all the information in real time, starting from the information about the container stuffing process (e.g., FCL—full container load consolidation of the cargo for shipping) at a logistic center based in the consigner's country, the information about shipping processes (with the consideration of the specificity of maritime transport—vessel monitoring provided in all the areas of the World Ocean) and ending with the information about temporary storage processes, consolidation at port terminals and logistic centers based in the consignee's country. To sum up, it should be stated that all the logistic operations in which containers participate during the transport-logistic process must be monitored under the container door-to-door delivery service. One of the most integrated and most advanced ECTS-class systems in maritime transport is the AVANTE system (AVANTE International Technology Inc. 2013).

The system is composed of several key functionalities, the most important of which are the following (see more in: Miler 2015, 42; Ahn 2005, 1719–1727; Miler 2019, 224–225):

- Monitoring of the loading unit integrity based on a multi-sensor system of container monitoring (covering all six container panels), which practically eliminates a possibility of false alarms.
- Monitoring of personnel work related to container transportation and security based on Transportation Worker Identification Card (TWIC). The card allows the interested party to identify not only the location of containers but also the location of employees responsible for security of containers in real time. Additionally, in an emergency situation, employees are able to raise a discreet alarm using the *panic* function provided by the card;
- Monitoring of all means of transport, including sea-going vessels. The system provides constant monitoring of vessels with the use of the GPS, GPRS and SATCOM systems, assuming that all detected anomalies affecting safety and security of containers and vessels must be transmitted to the monitoring center in real time.
- An efficient system for data display on devices specified by the user (including mobile devices, such as smartphones).
- A unique and doubled monitoring system dedicated to containers with high value goods (HVG).

The ECTS-class AVANTE system is equipped with sensors that are able to detect practically all the anomalies specified in container maritime transport, such as the following (see more in AVANTE International Technology Inc. [2021]; Miler [2015, 44]; Miler [2019, 224–225]):

- anomalies to the temperature, humidity and pressure specified for a particular container
- any types of shock resulting from hitting, sudden shifting, falling, etc.
- sound anomalies (characteristic sounds of sawing, metal burning, etc.)
- light intensity anomalies detected when the light inside a container is intensified (e.g., after an unauthorized opening of the container doors)

A general operating rule of the system is presented in Figure 11.2.

All anomalies, unauthorized openings of container doors, are detected by respective sensors (usually operating under the RFID technology) and immediately transmitted to the relevant (previously specified) institutions. The system operates on the exception principle—an *exception* is an anomaly from the specified standards and transportation regimes (NORMAL) detected by adequately configured subsystems (RELAYER, ZONER).

As mentioned, the ECTS AVANTE system is equipped with sensors that are able to detect specified anomalies and also with a communication subsystem that transmits the information about a threat to the specified addresses or provides remote access to the monitoring display on a particular website (after logging in with the use

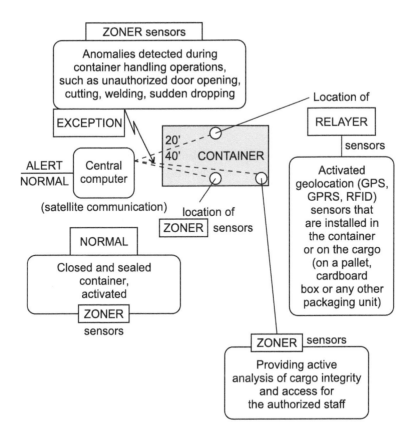

FIGURE 11.2.   A conceptual operating scheme of sensors applied in the ECTS AVANTE system.

Source: Authors' elaboration adapted and modified from AVANTE International Technology Inc. (2013).

of the login and the password assigned to the user). An example of the information about movement, location and status of a container in the logistic-shipping process in maritime transport is presented in Figure 11.3.

The system described previously comes as a representative example of capabilities and technical possibilities provided by electronic systems of container monitoring in maritime transport. Apart from undoubtedly important functions contributing to the improvement in security of shipping processes, similar systems provide their users with additional economic benefits, such as the following (cf. AVANTE International Technology Inc. 2013; Miler 2016a, 228; Miler 2015, 44; Miler 2019, 228–229):

- direct benefits for private companies, including:
  - improvement in the efficiency of logistic processes;
  - improvement in the reliability and competitiveness of the company; and
  - lower expenses spent on other security and safety systems;

- direct benefits for the public sector, including:
  - improvement in the efficiency of control processes implemented by the government institutions;
  - improvement in the maritime security of the state;
  - improvement in the level of security in logistic processes implemented by the state sector;
  - lower external expenses in maritime transport, including those related to ecology; and
  - lower congestion and congestion-related costs in maritime transport;
- indirect benefits for the maritime transport sector, including:
  - economies-of-scale benefits resulting from lower unit costs and an increased volume of containers handled in maritime transport;
  - improvement in the capabilities and capacity of maritime transport hubs; and
  - positive effects of support provided to auxiliary transport-related and port-related activities in maritime economy.

Shipping containers in logistic chains must take place under the control and monitoring regimes mentioned previously. However, the accuracy and optimization of loading and unloading procedures become significant issues in the discussed process. At present, managing these processes in container maritime transport is being taken over by systems of the vessel-planning class.

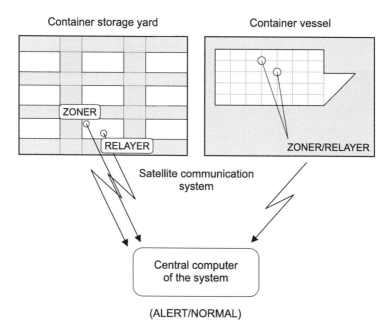

FIGURE 11.3 The AVANTE system for container positioning in maritime transport (a container terminal/a vessel).

Source: Authors' elaboration adapted and modified from AVANTE International Technology Inc. (2021).

11.3 LOADING SYSTEMS APPLIED IN MARITIME TRANSPORT OF CONTAINERIZED CARGO (VESSEL-PLANNING CLASS SYSTEMS)

Considering the high complexity of processes related to the management of cargo in container maritime transport, it is possible to define three basic levels of management, namely the following:

- the level of operational and logistic optimization in logistic chains:
  - on the vessel—optimization of the occupancy of slots and the use of bays, availability of containers for port and container handling operations through their optimal positioning in the stacks that can minimize the number of container movements; and
  - at the terminal—proper preparation of containers for loading operations (e.g., complying with the specified cut off time) and their convenient accessibility for RTG/STS cranes;
- the safety level:
  - on the vessel—optimization of vessel stability—metacentric height; and
  - at the terminal—the way containers are placed/separated on the terminal storage yards, with the consideration of the requirements stated in the Convention on the carriage of dangerous goods by sea (IMDG Code); and
- the stowage safety level of a container unit (stowage with the use of a lashing system, twist-locks, distribution of weight in a 40/60 container, in compliance with the stowage plan).

Considering issues related to the application of telematics, it is possible to state that the previously mentioned levels are characterized with diversified scopes of applying IT (telematics) processes and processes of physical handling of container loading units (that support digitalized processes of the information flow). A broader scope of digitalization (considering the information supply) can be observed at the level of operational and logistic optimization in logistic chains, when the slot exchange (mutual freighting of available container slots) takes place on the vessels that belong to an alliance. The lowest level of digitalization can be observed in processes involving physical container handling when containers are moved and stacked.

Presented previously, the range of problems and challenges related to management of cargo in maritime container transport has resulted in the necessity of implementing specialized software to handle processes, to support management and to provide process optimization. This kind of software is referred to as the vessel-planning-class software (a stowage plan developed to load the cargo). The SimpleStow software comes as a good example aI this point.

The SimpleStow is comprehensive software dedicated to the optimization of processes in container loading and exchange of information about containers (and their cargo) among the stakeholders of those processes—that is, parties that plan loading operations at the level of shipping companies, container terminals, maritime and government agencies (which are responsible for operations related to inspections and security of cargo). The software allows them to perform efficient loading, reporting and electronic exchange of information.

Despite its compact functional structure, the SimpleStow comes as the integrated software with the following functionalities (based on: AMT Marine 2021):

- SimpleStow systemic interface—to provide access to the functionalities of the system based on a purchased subscription
- Ship Model Editor—a graphical tool to visualize the loading space, depending on the type of a container vessel (considering diversified bay distribution)
- Web-based services—an Internet service to provide the exchange of files with individual models of loading space/vessel holds (models already existing in the database)

The SimpleStow uses modern graphical software to display a stowage plan for containers (with their cargo) with the possibility of obtaining a control printout. The application windows allow users to do the following (based on AMT Marine 2021):

- obtain a precise display of the container loading status for any selected bay, generated in accordance with the model of the vessel hold assumed for the particular type of a container vessel, with the consideration of the weight of containers and stability issues
- obtain a general display (an overall bay plan) showing the occupancy of all the bays and slots
- generate a list of containers (a list of cargo) in the form of a table (weight, cargo by IMO nomenclature, consignee)
- obtain a plan of the sailing (*Voyage Scenario, Port Rotation*) with the information about the subsequent cargo loading and unloading ports (POL—*Port of Load*, POD—*Port of Discharge*) in order to avoid loading conflicts during the entire sailing

The graphical interface of the SimpleStow software is presented in Figure 11.4.

The application provides the possibility of using a color code to arrange containers for loading and unloading operations planned at the subsequent POL and POD in a graphical way. The hold covers (their open/closed status), weight, container dimensions, stowage margins are also graphically presented to avoid problems related to overstowage. The display of the data in the Cargo List functionality allows the user to assign icons to the particular types and classes of containers, providing quick visualization of the containers undergoing handling processes.

Planning loading operations in the SimpleStow application (Cargo List/Booking List) provides the user with the possibilities of sorting, filtering, grouping, searching container lists, based on the following functionalities:

- graphical presentation of containers being handled
- cooperation of the graphical application with the databases in csv files

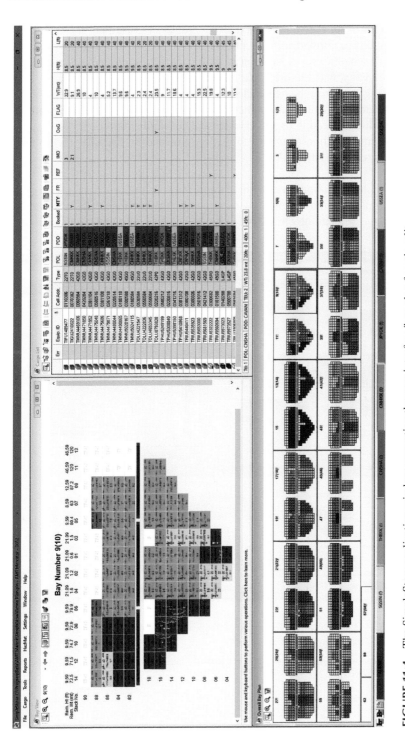

FIGURE 11.4   The SimpleStow application window presenting the user interface/bay plan functionality.

Source: Reprinted with permission of Vladimir Babakov, Managing Director (AMT Marine Software 2021, accessed: 12th April 2022).

- selection of an individual container or groups of containers for handling operations performed at POLs and PODs
- moving containers to a bay plan is done with the use of the drag-and-drop functionality, with the use of a computer mouse
- counting containers being handled in order to avoid mistakes in the process of assigning containers for particular operations at POLs and PODs, with a possibility of creating lists in csv files and generating reports
- cooperation with structured information provided by the EDI (Electronic Data Exchange) application

The Cargo List Functionality in the SimpleStow is presented in Figure 11.5. Planning a sailing in the SimpleStow application (*Voyage Scenario/Port Rotation*) allows the user to plan the entire route with the consideration of all the container

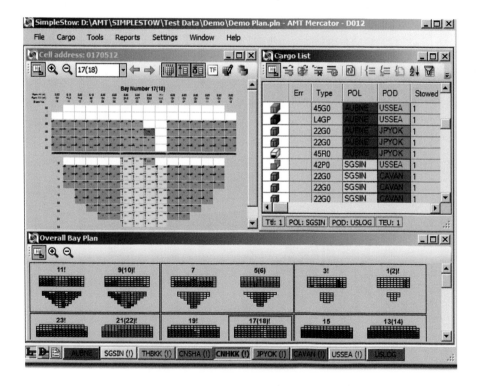

FIGURE 11.5 The interface of the SimpleStow application displaying the Cargo List/Booking List/Overall Bay Plan functionalities.

Source: Reprinted with permission of Vladimir Babakov, Managing Director (AMT Marine Software 2021, accessed: 12th April 2022).

Legend:

operations at all the POLs and PODs located along the route. It is possible with the use of the following functionalities:

- Each port is color-coded in order to improve the transparency of the bay plan (colors can be assigned by the system operator).
- A change in the location of containers at a particular port is automatically transferred to the subsequent bay plans for the subsequent POLs and PODs (there is no need to change the parameters of the ultimate overall stowage plan manually).
- During the entire sailing the system controls, analyzes and indicates potential problems with overstowage (*Stowage Conflicts Checking Routine*).
- Changes to the stowage plan can take place at any stage of the sailing and the system automatically recalculates the bay plan, the shipping route and other factors indispensable for proper operational processes, such as ETA (*Estimated Time of Arrival*) at NPOCs (*Next Ports of Call*).

The SimpleStow application also allows the user to choose the method of planning cargo handling operations with the use of two functionalities (AMT Marine Software 2021):

- *stowage from the cargo list*
- *direct stowage in the bay plan*

The *stowage from the cargo list* functionality is a tool recommended to provide a preliminary stowage plan. First, it is necessary to enter information about all the containers and their parameters (size, cargo, weight, POL and POD) manually or by importing csv files. Then, it is possible to indicate individual containers or groups of containers to be placed in a particular bay. On the other hand, using the *direct stowage in the bay plan* functionality, the operations are done in the reverse order: first, a container is placed in a particular slot in the bay plan (completing all the necessary container data required) and then the application automatically adds the particular container to the cargo list. This functionality is recommended when an ultimate plan is developed before the departure of the vessel.

The SimpleStow provides the user with an interface dedicated to the exchange of information in the EDI format, under the UN EDIFACT, BAPLIE 3.1 standards, responsible for data transfer to the customs and tax authorities and VERMAS which provides the user with the possibility to transfer data concerning container mass (VGM—Verified Gross Mass) to the port and terminal authorities.

Considering the fact that the SimpleStow requires the user to operate on various configuration models of vessel holds characteristic for the particular container vessels, the application provides a possibility to access the existing database on the structure (architecture) of bay plans or a possibility to generate the user's own model and to share it with other users providing them with the access from the application level. This can be done with the use of the *Model Editor* functionality and its graphical interfaces.

Information from the monitoring systems of container vessels and systems of cargo planning becomes input information for telematics systems of container maritime transport characterized by the highest integration capabilities—terminal operation systems.

11.4 TERMINAL OPERATION SYSTEMS DEDICATED TO MARITIME TRANSPORT OF CONTAINERIZED CARGO (THE TOS-CLASS)

Depending on the type and purpose of a port, each of its terminals (including container terminals as well) handles several or even more than ten vessels every day. Hence, cargo loading and unloading operations must be particularly efficient and precise. It is necessary to consider the full scope of the comprehensiveness of the process, namely a synchronized sequence of logistic operations performed in the area where maritime and land modes of transport meet (less often—inland waterway transport). The key aspect of timely and safe port and terminal operations is a possibility of precise planning such processes in advance, with the consideration of information provided in real time by the systems of monitoring and cargo planning, with the use of integrated solutions of the TOS (Terminal Operation System). Applying a TOS-class system in containerized cargo turnover that takes place at logistic facilities, such as maritime terminals ro-ro terminals, inland waterway or railway terminals, results in operational and economic benefits, including the following:

- optimization of management processes and their graphical display
- shorter time of container handling operations
- optimization of the use of cargo handling facilities
- shorter time of vessel handling operations/stay at berth

The main TOS functionalities (modules) are presented in Figure 11.6.

KPI	Resource planning	Reefer management	EDI	Document management
Vessel and berth planning	Rail planning and processing	Yard planning and optimization	Equipment dispatch	Gate planning and processing
Billing	Web	CFS	Advanced reporting	Customs compliance

FIGURE 11.6 The main TOS functionalities (modules).

Source: The authors' own elaboration.

Developed by Tide Works, a software company from the USA, The SPINNAKER is an example of a TOS-class system that can provide IT support to the process of container handling by simultaneous (Tideworks 2019):

- time optimization through a graphic display of a scenario for loading and unloading operations, with the consideration of capabilities and capacity of gantry cranes and other cargo handling devices;
- consideration of the sequence of containers which are arranged for handling at NPOCs, in accordance with the cargo manifest (cooperation with the vessel-planning class application); and
- consideration of vessel stability and proper metacentric height (cooperation with the vessel-planning class application).

The scope of cooperation between the TOS application and the vessel-planning application is presented in Figure 11.7.

The system allows the user to achieve numerous operational benefits and it significantly limits risks that can occur during operations that are performed, considering physical and security aspects. The benefits can be achieved through the following (Miler 2011, 216; Miler 2019, 214–215):

- High precision in operational planning with simultaneous maintenance of indispensable flexibility.
- The time period when the vessel stays at the berth is shortened to the minimum.
- Optimization in field of using one's own cargo handling facilities.
- Time of operation planning becomes shorter and potential human errors are eliminated.
- Current monitoring of threats that have been specified.

The *Spinnaker Planning Management System* tool (the managerial functionality of the TOS) allows the user to integrate and harmonize basic information that affects optimization of using the space and facilities of the terminal:

- parameters of the vessel and the cargo
- occupancy and parameters of the wharves, piers and quays
- the use of cargo handling facilities and storage yards
- availability and timing of the use of means of land transport (railway, container semi-trailers with tractors/maneuvering terminal tractors, etc.)

The Spinnaker Planning Management System® provides the following systemic tools (cf. Miler 2011, 218; Miler 2019, 216–217; Tideworks. 2019):

- Graphical Planning Tools—for intuitive management of the terminal space and the flow of containers from vessels to the wharves and to the means of land transport

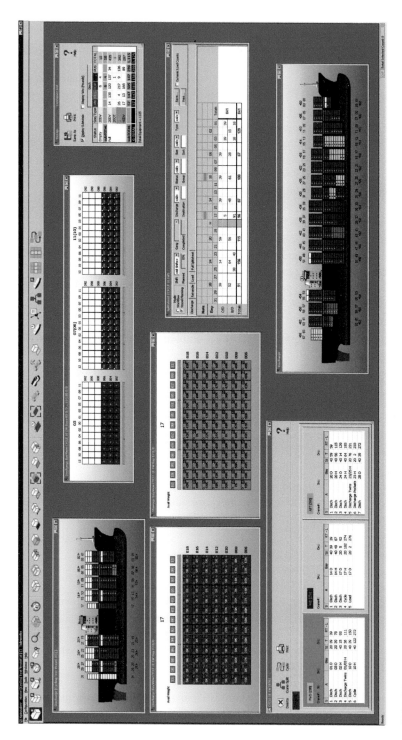

FIGURE 11.7 A window of the SPINNAKER application displaying the vessel-planning function.

Source: Reprinted with permission of Anthony Ricco, Business Development Manager, Tideworks Technology (Tideworks 2019, accessed: 24th May 2022).

- Vessel Berthing—to optimize the use of the wharves, with the consideration of their cargo handling capabilities and vessel size
- Yard Navigator—to use the storage and operational space of the terminal in the optimal way, which immediately allows operators to identify the location of stored containers and to minimize processes in which containers are moved around the terminal
- Vessel, Yard & Rail Editor—to define parameters and detailed characteristics of vessels, the resources of the container terminal and the capacity of railway tracks in order to optimize their mutual relations
- EDI (Electronic Data Interchange)—to exchange communications and documents under the EDI standard among the entities involved in the operations of a logistic chain
- Advanced Reporting—to streamline the flow of documents in customs procedures and inspections
- Administrator-Controlled User Rights Management and Authentication Capabilities—to provide management and administration of the system and to define procedures and access levels to various users of the system

Implemented to the SPINNAKER application, modern tools for 3D graphical visualization introduce much more user-friendly and more realistic functionalities that optimize the process of cargo planning, with the use of the following options (Tideworks 2021b):[1]

- Simulation modelling of the use of terminal space, resources (including suprastructural resources), equipment and processes implemented in real time; it allows the user to identify the problems, bottlenecks and anomalies; it results in higher productivity (the *full operational visibility* functionality).
- Selection of the particular processes for detailed visualization and improvement in managerial efficiency—e.g., vessel loading and unloading processes (the *vessel operations filter* functionality—Figure 11.8), processes of handling the railway terminal (that is an integral part of the container terminal—the *rail operations filter* functionality—Figure 11.9) or processes of using suprastructural equipment (the *handling equipment filter* functionality—Figure 11.10).
- Safety of storage and handling containers with hazardous cargo of the HAZMAT class (the *hazard filter* functionality—Figure 11.11).

To sum up, in order to perform their optimizing functions in terminal management, TOS-class systems, by a *de minimum* rule, should have capabilities to provide management of all the processes physically taking place in the area of a terminal in real time, to identify all types of bottlenecks and to optimize them (preferably in an automatic way, with the use of artificial intelligence elements included), to cooperate with vessel-planning systems and cargo handling systems applied in complementary modes of transport (railway, road, inland waterway under the IPSI), to optimize operation and use of the internal suprastructure of a terminal, to cooperate with ERP, EDI-class systems, etc.

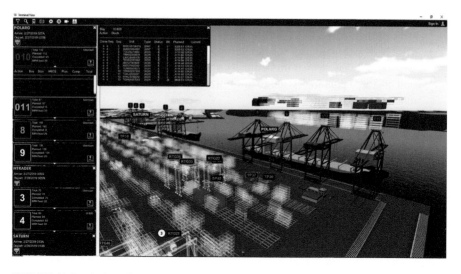

FIGURE 11.8   An interface of the 3D SPINNAKER application displaying the *vessel operations filter* functionality.

Source: Reprinted with permission of Anthony Ricco, Business Development Manager, Tideworks Technology (Tideworks 2021b).

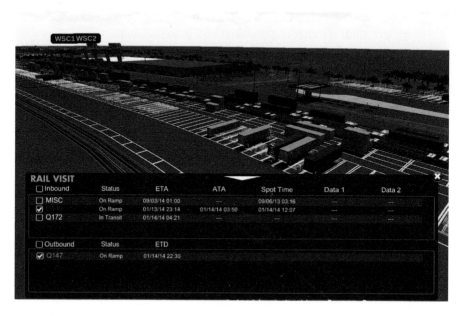

FIGURE 11.9   An interface of the 3D SPINNAKER application displaying the *rail operations filter* functionality.

Source: Reprinted with permission of Anthony Ricco, Business Development Manager, Tideworks Technology (Tideworks 2021b).

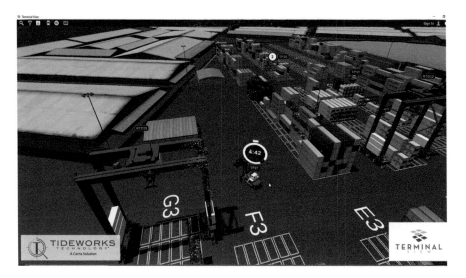

FIGURE 11.10   An interface of the 3D SPINNAKER application displaying the *handling equipment filter* functionality.

Source: Reprinted with permission of Anthony Ricco, Business Development Manager, Tideworks Technology (Tideworks 2022b, accessed: 24th May 2022).

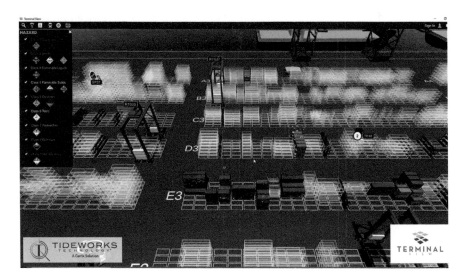

FIGURE 11.11   An interface of the 3D SPINNAKER application displaying the *hazard filter* functionality.

Source: Reprinted with permission of Anthony Ricco, Business Development Manager, Tideworks Technology (Tideworks 2022b, accessed: 24th May 2022).

The previously mentioned requirements are met by numerous applications offered as TOS-class software by various software development companies. The most popular applications of this type in the world are Navis, GullsEye, TBA—Autostore Terminal Operating System (TOS) and CommTrac (TOS), Contcloud as a unique SaaS (Software as a Service), HogiaTerminal, Mainsail, Octopi—CyberLogitec's OPUS Terminal, Realtime Business Solutions, TOPS Expert Cloud (Opus Terminal 2021).

However, none of the mentioned TOS-class systems would be able to optimize terminal operations without efficient access systems.

11.5   ACCESS SYSTEMS TO A TERMINAL OPERATION SYSTEM DEDICATED TO CARGO CONSIGNERS (ACCESS SYSTEMS FOR FORWARDERS, A NOTIFICATION SYSTEM, OCR)

The efficient operation of a terminal is achieved by processes involving synchronization of operations performed at the terminal yard, supported by the TOS system and in cooperation with entities who provide feeder services with the use of means of transport (forwarders, carriers) and who require an efficient, operational access system to the TOS (respecting the assigned authorization). The NavisSparcs N4 is an example of TOS (Navis) access functionalities provided to forwarders (it has been implemented by the DCT). The system allows forwarders to perform the main operations at choice, including the following (DCT 2022):

- *Unit*—identifying an individual container, a list of all the containers released for freight forwarding, an at-hand list of containers and bills of lading
- *Gate*—notification of containers in import and export and empty containers, bookings
- *Yard*—a list of additional orders
- *Cargo*—bills of lading, a list of notified general cargo submitted for shipping, a list of cargo in storage

By selecting the Unit tab, it is possible to display a list of all the containers and cargo released for a particular forwarder (as consignment for freight forwarding). The information includes the number of a container (Unit Nbr), the reception date (Time In) and discharge of a container/cargo (Time Out), the container operator (Line Op), a seal (Seal Nbr), the customs clearance number (CEN Number), a type of customs detention (Unit Impediments), the number of a summary declaration (DSK), a storage order assigned to a particular container (Service Order), the number of a storage order assigned by the system (Service Order Number) and many others. The tab also provides the user with access to the Unit Inspector card. The interface of the NavisSparcs N4 system displayed in the *Unit* tab is presented in Figure 11.12.

To facilitate the use of the system, the functionalities of the list can be arranged individually, according to the requirements specified by the forwarder (the selected items will appear in columns). An example presenting a personalized list of containers for the particular freight forwarding operation is provided as follows, in Figure 11.13.

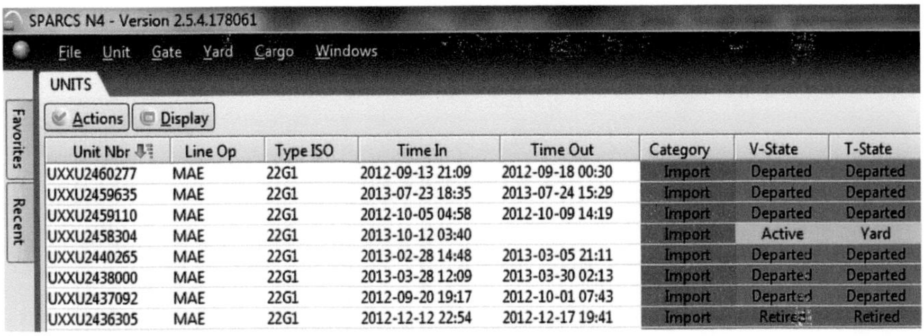

Unit Nbr	Line Op	Type ISO	Time In	Time Out	Category	V-State	T-State
UXXU2460277	MAE	22G1	2012-09-13 21:09	2012-09-18 00:30	Import	Departed	Departed
UXXU2459635	MAE	22G1	2013-07-23 18:35	2013-07-24 15:29	Import	Departed	Departed
UXXU2459110	MAE	22G1	2012-10-05 04:58	2012-10-09 14:19	Import	Departed	Departed
UXXU2458304	MAE	22G1	2013-10-12 03:40		Import	Active	Yard
UXXU2440265	MAE	22G1	2013-02-28 14:48	2013-03-05 21:11	Import	Departed	Departed
UXXU2438000	MAE	22G1	2013-03-28 12:09	2013-03-30 02:13	Import	Departed	Departed
UXXU2437092	MAE	22G1	2012-09-20 19:17	2012-10-01 07:43	Import	Departed	Departed
UXXU2436305	MAE	22G1	2012-12-12 22:54	2012-12-17 19:41	Import	Retired	Retired

FIGURE 11.12   A screenshot displaying the interface of the NavisSparcs N4 system (access for forwarders) in the *Unit* tab.

Source: Reprinted with permission of Kaczorowska Marta, Marketing Business Partner DCT Gdańsk, Poland (DCT 2022, 12).

The NavisSparcs N4 system also provides the user with a possibility of creating a personalized list of containers for a particular forwarder by selecting a group of containers on the list and adding them to the *My List* functionality. The display of *My List* is analogical to the display of the list in the *Unit* tab.

For efficient container handling (both by the TOS system and by the forwarder), the significant information is provided in the Unit Inspector card. The Unit Inspector is displayed only after the container is released for the particular forwarder. In the NavisSparcs N4 system, this functionality allows the user to indicate the container by entering its number into the system. After the particular container has been selected, the access to the functionality/the scope of information divided into the following sections is provided (DCT 2022):

- *Container*—the number of a container, the ISO type, the damage code (if applicable)
- *Status*—full/empty, the operator, weight, booking, storage orders submitted (Service Orders) and their assigned numbers (Service Order Numbers), the IMDG classification code (if applicable)
- *Transit*—the category (export/import, storage), POD, sailing (voyage), reception/discharge dates

A more detailed scope of information can be obtained with the use of the *detailed information* option/functionality, which provides an access to the following tabs:

- *AllEquipment*—the number of a container, the ISO type, the damage code (it refers to the container itself)
- *Contents*—a description of the cargo, net weight, details referring to the cargo of the IMDG and oversized goods categories (if applicable)
- *Damages*—the type, the location and the degree of the damage (minor, major)

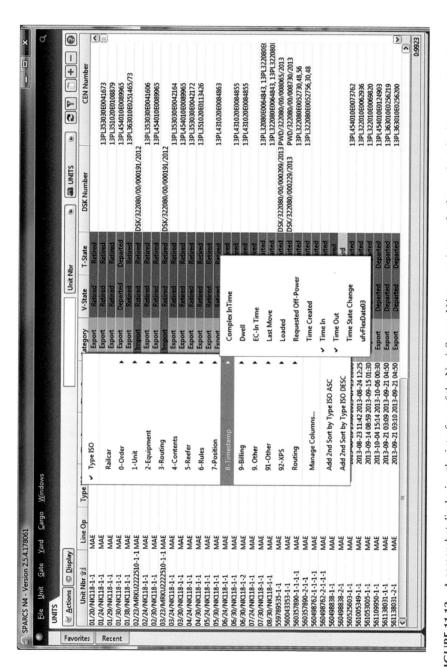

FIGURE 11.13 A screenshot displaying the interface of the NavisSparcs N4 system (access for forwarders) in the personalized *Unit* tab.

Source: Reprinted with permission of Kaczorowska Marta, Marketing Business Partner DCT Gdańsk, Poland (DCT 2022, 13).

- *Data Sources*—the source of information in the system:
  - *Declared Goods*—numbers of customs clearance operations
  - *History Move*—history of a particular container
  - *Holds/Perms*—holds and permissions imposed on or granted to a container
- *Itinerary*—the internal operational information of the TOS system
- *PrimaryEquip*—all the seals put on a container and general information about the container

An example of a Unit Inspector display (with the *AllEquipment* tab) is presented in Figure 11.14.

The efficiency and speed of container handling processes taking place at a terminal largely depend on the number of holds and permissions which require additional operations. Table 11.1 presents typical holds and permissions applied in container turnover at maritime terminals.

While using the *Gate* functionality in the NavisSparcs N4 system, it should be remembered that the right to enter the terminal is granted only to the containers and general cargo which have been previously notified in the TOS system (Navis). The *Gate* functionality is dedicated to export/import trade.

Export—notification of export containers is performed automatically through sending information about a particular booking by a shipping company, directly to the TOS (Navis) system. Import—if a particular container does not undergo any additional storage procedures and there is no necessity to view its Inspector Unit

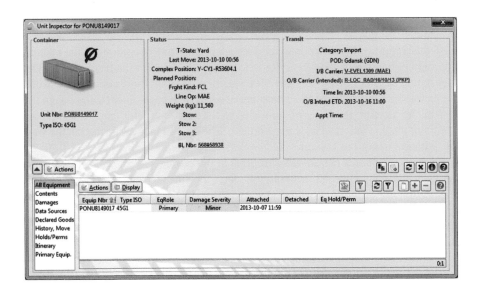

FIGURE 11.14   A display of a Unit Inspector (the *AllEquipment* tab active).

Source: Reprinted with permission of Kaczorowska Marta, Marketing Business Partner DCT Gdańsk, Poland (DCT 2022, 15).

TABLE 11.1

The List of Holds and Permissions Applied in Maritime Container Turnover

Name of a hold or a permission	Description
Holds	
Customs Export Hold	Applied in export by the Customs Office
Customs Import Holds	Applied in import by the Customs Office
IGDR/IZRX Hold	Applied by the Customs Office Risk Assessment Group
SL Hold (Shipping Line Hold)	Applied by the shipping company
Terminal Hold	Applied by the terminal
Tranzyt Karnet Tir	Applied by the Customs Office
Vet Hold	Applied by Veterinary Inspection
Sanepid Hold	Applied by the State Sanitary Inspectorate
WIORIN Hold	Applied by the Provincial Inspectorate for Plant and Seed Protection
WIJHARS Hold	Applied by the Provincial Inspectorate of Commercial Quality of Agricultural and Food Products
Permissions	
Customs Import Permission	The systemic overlay on all the import containers
Line Export Permission	The systemic overlay on all the export containers without the numbers of the export clearance entered into the system
Line Import Permission	The systemic overlay on all the import containers that have not been released by the shipping company; the release takes place when the terminal obtains the PIN from the shipping company

Source: The authors' own elaboration based on DCT (2022, 15).

card, the shipping of such a container does not require any operations in the system. The only obligation of the forwarder is to provide the PIN to the truck driver. The PIN should be submitted to *Pregate*. The situation becomes more complicated when the forwarder's code (*Agent PIN*) must be assigned to the container (*Unit Nbr*) in the system. As a result, the access is provided to the Inspector Unit card with the possibility of placing orders (e.g., storage orders). The operation is performed in the *Unit* tab, with the use of the *Validate PIN for Self-Assignment* functionality (DCT 2022).

In the NavisSparcs N4, in order to provide the customs agent with an access to the Inspector Unit card, it is necessary to enter the customs agent's code (*Shipment Details Agent*). After this operation, in the Agent One box the forwarder's code appears (e.g., NK001), and in the Agent Two box, the customs agent's code is displayed (e.g., AC001).[2] To perform actions such as inspections, container stuffing/stripping, container weighing, etc., it is necessary to place and order (under CFS) in the TOS (Navis) system. Placing orders (*Create Service Order*) in the *Transactions* functionality is possible (DCT 2022):

- in export—only after the previous assignment of the freight forwarding code to the particular container by the shipping company
- in import—after the assignment of the PIN to the particular container by the forwarder

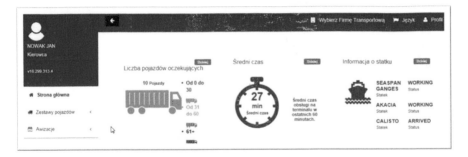

FIGURE 11.15   The interface of the main window of the e-BRAMA application.

Source: Reprinted with permission of Kaczorowska Marta, Marketing Business Partner DCT Gdańsk, Poland (DCT 2021, 6).

Another system that cooperates with the TOS is a notification system, understood as a tool dedicated to truck drivers, shipping companies and forwarders to provide them with a possibility to plan their arrivals at the terminal so that the time (arrival/departure) parameters are optimal for all the parties and the entire logistic chain is streamlined. An IT solution that cooperates (is synchronized) with the TOS NavisSparcs N4 is the e-BRAMA application (at the DTC—as an example discussed here). After logging into the system, basic information about terminal operations is displayed, including the number of waiting vehicles, the average time of handling operations, information about vessels and tabs with the selection of notifications and vehicles (Figure 11.15).

The notification links the driver, the truck and the semitrailer with containers, booking a time slot (usually 2 h), when the driver is going to be handled at the DCT. The system allows the user to create notifications only for the containers which are ready for discharging or receiving. With some exceptions, it is impossible to notify a container with a hold or in a situation when the mass of the container, the truck and the semitrailer exceeds the permissible total weight.[3]

To create a notification in the e-BRAMA system, it is necessary to do the following:

- Add the container to the list of tasks (and, in this way, to verify the basic data about the discharged or received container—the PIN for an import container or the booking number for an export container).
- Link the container with the vehicle that is going to transport it.
- Define the time slot (the slot + buffer, usually 1.5h).

These operations allow the user to notify containers properly. In the e-BRAMA system, the commencement of the notification procedure is displayed in the respective tab (Figure 11.16).

Usually, a notification is possible after the seal numbers are provided.[4] It is necessary to state at least the shipowner's seal (the operator's seal). If the seal number is not provided, the system displays a hold (a graphical symbol of a red padlock). The interface of the e-BRAMA system with the information about holds is presented in Figure 11.17.

FIGURE 11.16 The notification tab in the e-BRAMA system.

Source: Reprinted with permission of Kaczorowska Marta, Marketing Business Partner DCT Gdańsk, Poland (DCT 2021, 8).

FIGURE 11.17 The interface of the e-BRAMA system with the information about holds.

Source: Reprinted with permission of Kaczorowska Marta, Marketing Business Partner DCT Gdańsk, Poland (DCT 2021, 10).

Legend:
- Permissions—granted automatically by the system; Required—a hold must be cancelled; Granted—a hold is cancelled; cancelled—cancelling a hold with the Granted status.
- Holds—are imposed manually, based on individual requirements; Active—it is necessary to cancel a hold; Released—a hold is cancelled.
- The number of a task is assigned automitically when the task is added.
- The number of the assigned notification—if a container is linked with the notification, the number of the assigned notification is displayed; 0 means that there is no link to the notification.
- Notices—additional information about whether the container is at the terminal, whether there are any holds imposed on it, for example, the MSCU 51050021 container with a notice; Yard and Stop means that the container is at the terminal and there is a customs hold imposed on it; a green padlock means that the seal number has been entered into the system, an open red padlock means that the seal number has not been entered yet.

To notify an export container in the e-BRAMA system, a prior notification must be made by the shipowner in the TOS Navis system. For import containers, it is possible to add containers to the list of tasks even before they are unloaded from the vessel and also when there are customs holds imposed on them. Finally, notification

FIGURE 11.18 The final confirmation of a notification in the e-BRAMA system.

Source: Reprinted with permission of Kaczorowska Marta, Marketing Business Partner DCT Gdańsk, Poland (DCT 2021, 12).

at the DCT terminal made with the use of the e-BRAMA (the last step displayed in the window presented in Figure 11.18) involves selecting containers and assigning them to the particular vehicles (with the use of the drag-and-drop function), selecting convenient time slots (the calendar and time slots with hours) and confirming the entire operation.

Automation of terminal processes is possible owing to ORC (*Optical Character Recognition*) systems that can automatically recognize signs (such as license plates, seals, container numbers, drivers' biometrical features from drivers' ID cards, etc.). OCR systems are installed at most maritime container terminals in order to optimize administrative procedures related to the arrivals of vehicles and their movements around terminal yards.

An example of an OCR system applied at maritime container terminals is the OCR system implemented at the DCT that is integrated with the e-BRAMA and TOS Navis systems. Both gates of the DCT, the entry gate and the exit gate, are equipped with two OCR portals and one OOG[5] each.

The portals must be passed with the speed of about 10km/h.[6] The license plates of vehicles and semitrailers should be clean to allow the system to read them properly. After the vehicle recognition and data confirmation, the vehicle can enter the terminal—the portal signals that with green lights. Vehicles must keep proper distance between one another to allow the system to read the data properly. If the data are not read in the proper way, it is necessary for a vehicle to turn around and to pass the OCR IN portal once again. The stages of proper entering the premises of the DTC maritime container terminal with the use of the OCR system form the following sequence:

- OCR IN
- Self-Service
- Gate IN
- Yard (physical container discharging or receiving)
- OCR OUT
- Gate OUT

In order to pass the subsequent stages—namely Self-Service, Gate IN, Gate OUT—drivers are required to have their Driver's Cards, which must be submitted at the Pre-Gate office personally by each driver.[7] Before approaching the Self-Service station, the driver must prepare their Driver's Card and the notification number (TVA/Booking/EDO).[8] At the Self-Service station, the data in the Driver's Card are verified by a proximity reader. After the positive verification of the data by the system, the gate barrier is opened automatically. If there are any anomalies detected, a systemic alarm is triggered. Such situations most often refer to Border Guard alarms (usually at the OCR OUT), incorrect reading or the lack of seals and incorrect reading or the lack of relevant IMDG markings (at the OCR IN). While passing the OCR portals, the system scans containers and detects seals. Therefore, seals must be put on containers before vehicles pass the first OCR portals. If the reading of the OCR IN portal is erroneous (the seal is put on a container), the employee at the entrance gate (Gate IN) makes a respective correction and allows the vehicle to enter the premises. If there is no seal on a container, the system detects its lack and the SEAL MISSING hold is imposed on the container in the system. The Pre-Gate informs the shipowner about the necessity of submitting an order of putting the seal on a container (under CFS, putting the seal).

Containers must be properly marked with labels informing about the type of dangerous goods transported inside (IMDG) before passing the first OCR IN portals. If the container marking is incorrect or there are no labels at all, the driver can still enter the terminal premises. However, a PLACARDS MISMATCH hold is automatically imposed on such containers and they cannot be loaded onto a vessel. In such a situation, based on the photos provided by the OCR, the Pre-Gate informs the shipowner about the discrepancies and requests to put the container in question under the CFS (completion or removal of incorrect labels). After the operation, the PLACKARDS MISMATCH hold is cancelled by the CFS.

If during the passing through the OCR OUT portal a Border Guard alarm is triggered, the vehicle must pull over to a specially prepared and marked parking place (RPM) and wait for the Border Guard officers. The vehicle can leave the premises of the terminal only after the consent of the Border Guard is granted.

Discussed in this chapter, all the functionalities of telematics systems applied in containerized cargo turnover in maritime transport and operations performed at container terminals clearly indicate the role and significance of telematics systems in the operationalization of processes, their optimization and improvement in management. It concludes the considerations of Part 3 dedicated to Managerial and Operational Challenges to Maritime Containerized Transport, with emphasis on entire *spectrum* of aspects related to optimization through the implementation of best managerial processes and IT (telematics) solutions/tools.

Taking all the determinants mentioned previously into account, one more aspect of growing importance is to be noted in order to finally conclude the holistic approach toward the pragmatics of containerization processes in maritime transport—namely sustainable development.

NOTES

1. Elaborated on the basis of Tideworks (2021b).
2. If the forwarder is also the customs agent and wishes to clear the container by following the simplified procedures, it is necessary to fill in the Agent Two box in the system as well (e.g., AC001) to provide access to the Inspector Unit card to the user who logs in on the customs agent's account.
3. Permissible total weight is the total weighr of the vehicle (or the group of vehicles) waiting and ready to go, with the cargo weight declared as permissible by the relevant authorities of the country where the vehicle is registered; in Poland, in accordance with the Law on Road Traffic, it is the highest weight of a vehicle loaded with cargo and passengers, defined by the relevant technical specifications, that is allowed to use roads.
4. Export containers can be entered into the system without their seal numbers (it is possible to notify them), but the seals must be added at the moment of slot commencement at the latest, by editing the task. While notifying flat-rack containers that do not have any place where seals could be fixed, it is necessary to enter NO SEAL information.
5. Aut. Out of Gauge—described in the previous chapters, the way of referring to all oversized cargo.
6. If a vehicle moves faster, there is a risk of erroneous scanning, which hinders or prevents the subsequent stages of entering the DCT premises.
7. The card can be issued exclusively for a driver who is registered in the e-BRAMA system (biometric data). The first Driver's Card is free of charge; each subsequent one is PLN40.
8. TVA—*Truck Visit Appointment*—is the notification number generated by the e-BRAMA system and sent as a short text message to the driver; EDO is the number assigned by the ship owner, indispensable to take empty containers.

Part 4

Sustainable Development Challenges to Containerized Maritime Transport

12 Ecologistics and Sustainable Development Requirements in the Pragmatics of Containerization Processes in Maritime Transport

12.1 THE ORIGIN AND FORMAL AND LEGAL ASPECTS OF EMISSION RESTRICTIONS IN MARITIME TRANSPORT

12.1.1 MARITIME TRANSPORT AS A GHG EMITTER—AN IMPERATIVE FOR THE IMPLEMENTATION OF A METHODOLOGY FOR COUNTING EXTERNALITIES AND EXTERNAL COSTS IN MARITIME TRANSPORT (INCLUDING CONTAINERIZED MARITIME TRANSPORT) IN THE LIGHT OF CHALLENGES POSED TO SUSTAINABLE DEVELOPMENT

Problems related to sustainable development are not totally new issues to global economy In the second half of the 20th century, the international community started to notice problems concerning overexploitation of the natural environment, although renewable and nonrenewable natural resources were still estimated without the consideration of real external costs of the natural environment. Also, some advanced processes related to the internationalization of the global economy appeared, and this fact was translated into an increased demand for all types of raw materials and goods. Globalization also resulted in a considerable increase in transportation needs that would allow interested parties to maintain the optimal character of logistic supply chains. The dynamic development of international trade and intensification of shipping processes (including shipping containerized cargo by sea) resulted in the serious pollution of natural environment (understood as an impact of the anthropopressure factor of transport). The awareness of a degradative role of economic activities, including shipping operations, on the natural environment sparked a number of international discussions and generated numerous documents (including general legislative acts) aimed at the implementation of more ecological solutions—for example,

DOI: 10.1201/9781003330127-17

a report of the Secretary-General of the United Nations, Mr. U Thant, *Problems of human environment* (United Nations 1969); the final document of the United Nations Conference on the Human Environment—UNCHE in Stockholm (Nicholls 1973, 117–119);[1] an agency of the United Nations Environment Programme UNEP as an outcome of the United Nations Conference in Stockholm in 1972; and an outcome of the sessions of the Club of Rome (1972–1976), a concept of sustainable development, understood as balance between the opposing forces of economic growth and the reduction of pollution (Meadows, Randers and Behrens III 1972, 171).

Hence, the essence of sustainable development is to maintain the natural balance and integrity of natural processes, with the consideration of economic growth at the same time (Prawo ochrony środowiska 2001; European Economic and Social Committee 2020). Furthermore, in accordance with the objectives, sustainable development should be based on integrated actions pertaining to various fields of human activities, from economy to politics, where three fields of activities were initially the most significant ones (Dąbrowski 2013, 28):

- ecological—preserving natural environment possibly in an unchanged form and counteracting its deterioration
- social—adopting the principle of intergenerational justice and responsibility for natural environment
- economic—self-restraint of societies in their exploitation of natural resources and implementation of methods for recovering such resources

Further actions undertaken by the United Nations were related to setting objectives for sustainable development. In 1992, the United Nations Conference on Environment and Development (UNCED) in Rio de Janeiro adopted the *Global Action Programme—Agenda 21*. In 2012, another United Nations Conference was held in Rio de Janeiro (United Nations Conference on Sustainable Development, Rio+20—UNCSD), which confirmed further implementation of the sustainable development policy. Additionally, a funding strategy for sustainable development was developed and implemented, and also, the significance of further implementation of ecological regulations and concepts of sustainable development in a broadly understood social field was indicated (United Nations 2012). In 2015, the General Assembly of the United Nations adopted a resolution *Transforming our World: The 2030 Agenda for Sustainable Development*, which came as the updating to the objectives of sustainable development for the subsequent years (United Nations 2015). Among the 17 postulates, social conditions and activities in the field pertaining to protection of natural environment are particularly significant. Counteracting the effects of climatic changes, protection of land and marine ecosystems and rational use of natural resources are also emphasized in the document.

The principle of sustainable development was also considered by the European Union in *Treaty on the European Union—TUE* of 1992. In the treaty (Art. 130), it is clearly indicated that sustainable development defines the main line for economic and social development of the European Community countries (Hartman 2009, 94). In the document, the parties adopted the objective on close cooperation among the EC member countries, including all the activities pertaining to protection of natural environment. When Eastern European countries joined the European Union in 1997, a new *Treaty of Amsterdam* was adopted, amending the *Treaty of the European Union*. The principle of sustainable development was included in the new document as a guideline for all the member countries (Official Journal of the European Communities 1997). In 2001, the European Commission published another document: *A Sustainable Europe for a Better World: A European Union Strategy for Sustainable Development* (Commission of the European Communities 2001a).

In the strategy, the postulates on undertaking efficient actions under sectoral policies of the European Union were adopted (Commission of the European Communities 2001b). Among other significant fields of activities, the policies pertaining to the following areas were indicated: agriculture, fisheries, development of regions and transport. The policy of sustainable development pursued by the European Union was maintained in a new strategy implemented in 2001 by the following document: *Europe 2020: A Strategy for Smart, Sustainable and Inclusive Growth* (Commission of the European Communities 2010).

In 2019, the implementation of the sustainable development strategy of the European Union was extended by 2030. The main guidelines are presented in *Reflection Paper Towards a Sustainable Europe by 2030*, which refers to the regulations included in the Agenda 2030 adopted by the United Nations (Commission of the European Communities 2019). Apart from achieving its main aim, namely sustainable development, the document indicates a model of circular economy in which waste is significantly reduced (Figure 12.1).

The policy of the European Union is closely related to the implementation of the objectives of sustainable development. An area that comes as a significant field of activities undergoing regulations of the EU policy is transport.

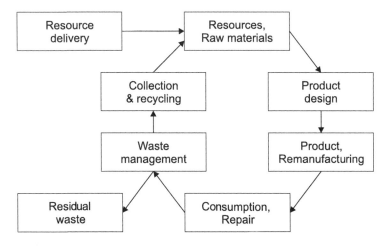

FIGURE 12.1   Circular economy—the EU development strategy until 2030.

Source: The authors' own elaboration based on Commission of the European Communities (2019, 15).

A condition for the efficient implementation of the principles for sustainable development is providing an integral transport policy that includes all the modes of transport.

In 1992, the European Union published a set of regulations on transport policy, *Green Paper—The Impact of Transport on the Environment*, where the impact of transport on natural environment is discussed. In the document, the main limitations to the development of transport are discussed, and the necessity to implement the objectives of sustainable development is indicated (Commission of the European Communities 1992b, 11–15). The main threats posed by transport to natural environment are sea and sea pollution, noise, the quality of transport functioning (congestion), risk related to transport of hazardous cargo and the impact of transport on spatial development (including reduction of agricultural and green areas). At the end of 1992, *White Paper on the Future Development of the Common Transport Policy* was published, in which the principles for a model of sustainable mobility are defined (Figure 12.2).

The concept of sustainable mobility assumes coexistence of two fundamental aims for the maintenance of an efficient transport market that offers competitive services and limits pollution of natural environment.

During the subsequent years, another document was developed and eventually issued in 1995: *Green Paper—Towards Fair and Efficient Pricing in Transport*. It presents problems concerning mechanisms that shape prices for shipping services with the consideration of external costs (Commission of the European Communities 1992a, 2). Another element of the EU transport policy is *White Paper—Fair Payment for Infrastructure Use: A Phased Approach to a Common Transport Infrastructure Charging Framework in the EU*. The White Book of 1998 indicates the necessity to provide a systemic solution for the structure of costs in shipping processes, with the consideration of external fixed and variable costs (Commission of the European Communities 1998).

Other elements of the transport policy of sustainable development are listed in the White Book of 2001. The *European Transport Policy for 2010: Time to Decide*

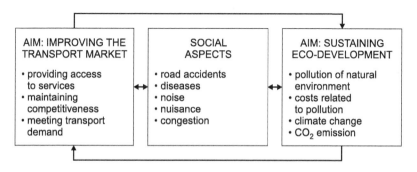

FIGURE 12.2 A model of sustainable mobility.

Source: The authors' own elaboration based on White Paper on the Future Development of the Common Transport Policy (1993).

is a document in which the implementation of systemic shipping processes is suggested along with further development of ecological modes of transport. In the document, the necessity for more sustainable use of the particular modes of transport is indicated, along with the need to use all the advantages of the particular transport systems. The significant activities that will allow interested parties to achieve sustainable development involve the following:

- achieving balance among the particular modes of transport by network development of transport systems, concepts of sea highways, short sea shipping and TEN-T transport corridors
- eliminating bottlenecks and congestions in transport systems by using multimodal transport, high-speed transport and implementing comprehensive infrastructural projects
- improving road safety and defining rights and obligations for users of transport systems; harmonizing regulations concerning payment for the use of transport infrastructure, including natural environment costs (internalization of external costs of transport)

A document that presents transport policy pursued by the European Union is *Keep Europe Moving—Sustainable Mobility of Our Continent*, a Communication of the Commission to the Council and European Parliament, published in 2006. The document states the policy objectives, such as development of efficient and effective transport systems that provide technological innovation and protection of natural environment (The Council and the European Parliament 2006, 3–4). A significant difference that can be observed in relation to the previous visions for sustainable development in EU transport is a systemic and complementary inclusion of all the elements of the transport system that can provide sustainable development. The complementary elements include the following: infrastructure, economic conditions in the markets of shipping services and standards for protection of natural environment.

Another document is *A Sustainable Future for Transport: Towards an Integrated, Technology Led and User Friendly System*, a Communication of the European Commission published in 2009 (Commission of the European Communities 2009).

In 2010 the European Parliament adopted a resolution, *A Sustainable Future for Transport* (Rezolucja Parlamentu Europejskiego 2010), where the objectives for further development of the EU transport policy are defined along with an indication of problems related to the growing emission of CO_2, SO_x, NO_x in maritime transport, a necessity of developing sustainable shipping and logistic chains based on decarbonized transport, under the European Transport Network (TEN-T).

A comprehensive document that defines lines for the development of transport in the European Union is the White Book of 2011, *Roadmap to a Single European Transport Area—Towards a Competitive and Resource Efficient Transport System* (European Commission 2011, 6–8).

Over the years, an evolution of the EU strategy for transport development has been observed, aiming at the improvement in the efficiency and availability of transport markets and reduction of natural environment pollution. By 2030, three specific objectives are going to be achieved (European Commission 2011, 9–11):

- developing and implementing new fuels and propulsion systems compliant with the principle of sustainable development
- optimizing operation of multimodal logistic chains
- improving efficiency of transport and developing infrastructure with the use of information systems and market incentives

According to the EU policy, CO_2 reduction in road transport is going to take place, and in urban areas, CO_2-emission-free zones are going to be implemented by 2030. Considering air and maritime modes of transport, it is necessary to introduce low-emission fuels, and their share should be 40% at least by 2050 (or even up to 50% if possible). In the field of the protection of the natural environment, a polluter-pays principle must be implemented. It will result in real costs from the transmission of pollution onto perpetrators of damage who have polluted the natural environment (assenting external costs of transport in the system of cost calculation).

Considering the concept of sustainable transport defined in this way, it should be noted that it also broadly refers to maritime transport (including containerized cargo). Sustainable shipping is a capacious international concept that has been developed by the IMO. It involves three practical requirements: no casualties, no pollution and service on time (Potts 2018, 96–97). The concept has been also developed by the European Maritime Safety Agency (EMSA), which defines sustainable shipping as "the overall concept of management (a holistic management concept) for sustainable development applied in maritime shipping that comprises responsibility for natural environment and social responsibility" (European Maritime Safety Agency 2022).

Although international shipping is said to be the most energy-efficient mode of mass transportation, only moderately contributing to the overall GHG (CO_2) emissions, a global approach toward further improvement in its energy efficiency and effective emission control is needed, especially with the consideration of the fact that sea transport facilitating seaborne trade (90% of worldwide intercontinental trade is facilitated by the maritime transport) will continue to grow (Brzozowska and Miler 2017, 195–209).

Carbon dioxide (CO_2) is the most important GHG emitted by the shipping industry in both terms: its total amount and its potential impact on global warming processes. Emission of CO_2 from shipping in comparison to the total (global) emission is depicted in Figure 12.3.

The level of CO_2 emission is related to the shapes of vessels, the dimensions of hulls, the efficiency of engines and the overall (including navigation, weather conditions, load etc.) performance of ships. The potential emission of CO_2 from various types of ships (including container ships) is depicted in Figure 12.4.

Despite the fact that the entire GHG emission from shipping industry is estimated for 1 billion tonnes a year and accounts for approximately 3% of the world's total (global) emission and 4% of the EU's emission, without any action undertaken, these

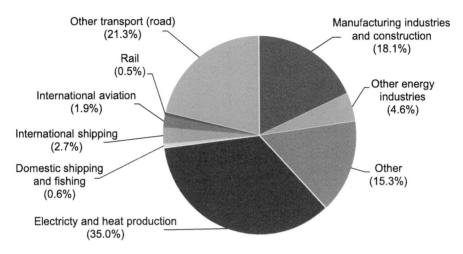

FIGURE 12.3 CO_2 emission from shipping in comparison to the total global emission.

Source: Based on Akoel and Miler (2019), Second IMO GHG Study (2009).

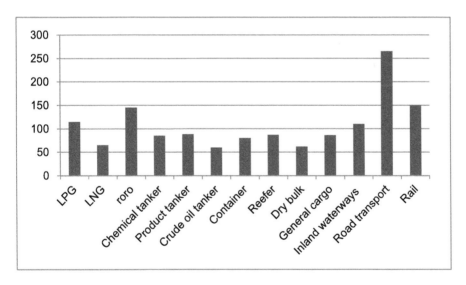

FIGURE 12.4   Comparison of emission levels generated by various types of 10–15,000 dwt vessels (gram of CO_2 per tonkm range per vessel type).

Source: The authors' own elaboration based on Jurdziński (2012b).

emissions are expected to rise and to be doubled by 2050 (Jurdziński 2012a). This remains in contradiction to an internationally agreed decision of keeping global warming below +2°C, which requires focusing on a decrease in emission worldwide (50% of the levels from 1990 by 2050) (IMO 2021). Figure 12.5 shows the previously mentioned estimations.

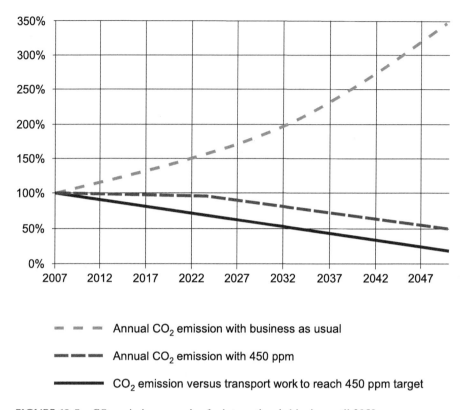

FIGURE 12.5   CO$_2$ emission scenarios for international shipping until 2050.

Source: Based on Akoel and Miler (2019), IMO (2021).

The results of these analyses have become a starting point for reducing negative estimates.

The data that are necessary for the calculation of emissions from ships while in transit and at ports may come from many different sources, such as the following (Akoel and Miler 2019; IMO 2021):

- data from port authorities and pilot station-data used for the calculation of the time of an individual ship
- data from the AIS-Automatic Identification System used for automatic exchange of data useful to avoid possible collisions between vessels and identifying ships for different onshore systems which monitor the movement of ships (VTS)
- data from the ship register with a detailed description of the vessel, providing information about the power of the vessel main and auxiliary engines
- ship daily reports should be considered as the most reliable source of information, particularly on the main and auxiliary engines, fuel consumption and volume of cargo (load)

All these vital prerequisites for environmentally friendly shipping demand integration and common (global) acceptance; thus, the EU and IMO have decided to introduce unified regulations in this field.

12.1.2 LEGAL ACTS AND LEGISLATION AND OPERATIONAL FACTORS (MARPOL, SOLAS, BWM, ENERGY EFFICIENCY DESIGN INDEX (EEDI), SHIP ENERGY EFFICIENCY MANAGEMENT PLAN (SEEMP)

The principles of sustainable development have been also implemented into the functioning of global maritime transport. The fundamental legal regulation that refers to maritime shipping and port operation is the International Convention for the Prevention of Pollution from Ships (MARPOL) (International Maritime Organization 2017).

The convention implements the guidelines for shipping particular types of cargo and procedures for vessel operation, aiming at reduction of marine environment pollution. The regulations also apply to management of oil waste, sewage, garbage and to prevention of air pollution generated by ships. Shipowners are obligated to implement the procedures stated in the Convention which refer to transportation of specified groups of cargo. In order to reduce sea water pollution, seaports are required to collect hazardous substances from ships and to utilize them on land. This requirement considerably decreases the amounts of liquid (mainly oil substances) and solid (plastic and non-organic materials) waste that cause permanent pollution of natural environment.

Initially, MARPOL was supposed to be a continuation of the legislation on requirements referring to shipping oil and oil derivatives by sea. Previous regulations were defined in the OILPOL conventions (International Conventions for the Prevention of Pollution of the Sea by Oil) which were subsequently adopted in 1954, 1962, 1969 and 1971. It should be noted that for many years discharges of oil-derived substances from ships and accidents of tankers had been considered as the main threat to marine environment. In 1983, there were two annexes added to the MARIOL Convention, namely Annex I—*Regulations for the Prevention of Pollution by Oil* and Annex II—*Regulations for the Control of Pollution by Noxious Liquid Substances in Bulk.*

Annex 1 provides a comprehensive regulation to the problems referring mainly to oil pollutants. Oil pollution (including oily water) must be kept in oil residue (sludge) tanks of the adequate handling systems and collected from vessels to the proper port installations. Tankers used for shipping crude oil and oil-derived substances and vessels that need liquid fuel for their propulsion systems must carry oil record books (International Maritime Organization 2012). An oil record book is a document which comes as an attachment to the log book of a vessel and it is used for recording all the operations involving oil and oil-derivatives undertaken on a vessel. The captain of the vessel or a person appointed by the captain is responsible for exercising full control over the following:

- loading and unloading oil substances from port terminals
- cleaning the vessel holds with the use of the crude oil washing system (COW), which involves washing holds with pressurized oil instead of pressurized water (International Maritime Organization 2000)

- ship bunkering
- managing any waste generated during vessel operation, including residues of de-oiled bilge water, residues of filtered fuel and lubricating oils and other residues from the vessel installations

Stated in the convention, the regulations refer to the construction of tanker vessels used for transporting oil and oil-derived substances. Tankers should have double bottoms and double hulls because this type of structure can protect oil tanks during collisions or running aground (International Maritime Organization 1996; Polish Register of Shipping PRS 2020, 5). The convention also defines special areas that require special protection of natural environment. Special areas usually comprise sea areas that are particularly vulnerable to pollution because of low levels of water exchange and low absorption of noxious substances. These areas include the Mediterranean Sea, the Baltic Sea, the Black Sea, the Red Sea, the North Sea, the Irish Sea, the Celtic Sea, the Caribbean Sea, the Gulf of Aden, the Persian Gulf, the Gulf of Mexico, the Antarctic, waters of South Africa. Considering the condition of natural environment, a discharge of any oil-derived substances is forbidden. Annex II to the convention provides a classification of liquid noxious substances that—when discharged into marine environment—pose a threat to marine resources, human health and sea amenities. These substances fall into four classes depending on the level of their harmfulness (International Maritime Organization 1978):

- X—noxious substances that present a major hazard to marine environment, and therefore, their discharge into sea water is strictly prohibited.
- Y—noxious substances that, if discharged into sea water, present a hazard to marine environment; therefore, there can be some limitations set to their discharge into the sea.
- Z—substances that present a minor hazard to marine environment, and they undergo less stringent restrictions referring to the conditions of their use.
- OS—other substances (not classified as X, Y, Z); they do not affect marine environment in a negative way.

The MARPOL Convention was successively amended with the subsequent annexes providing comprehensive regulations on operation of cargo vessels and procedures for shipping various cargo groups. In the subsequent years, the following three parts were added to the convention:

- Annex III—*Prevention of pollution by harmful substances carried by sea in packaged form* (in 1992)
- Annex V—*Pollution by garbage from ships* (in 1998)
- Annex IV—*Pollution by sewage from ships* (in 2003)
- Annex VI—*Prevention of air pollution from ships* (in 2005)

Annex III of the MARPOL Convention defines methods of preventing pollution of sea water with noxious substances carried by sea in packaged forms (International Maritime Organization 1992). Packaged cargo should be precisely identified by stating its technical name and labelling it with adequate hazard warning signs. Shipping noxious substances in packaging should be accompanied by adequate shipping documents, including a hazardous cargo declaration, packaging labels, a hazardous cargo manifest, a stowage plan of hazardous cargo. Listing noxious substances that pose threats to marine environment in Annex III results in the necessity of implementing shipping processes in accordance with the principles stated in the International Maritime Dangerous code (IMDG). The code provides a set of procedures to be applied in shipping particular cargo transported in unitized forms (packaging units, containers) (ICHCA International 2020).

Another group of pollutants defined in the Convention refers to waste (Annex V). Waste can be treated and discharged from a vessel if it meets standards of cleanness defined by the regulations for the conventional survey of seagoing vessels (Polish Register of Shipping PRS 2022, 33).[2] The prevention of sea pollution with waste (mainly plastic and non-organic waste) means that each vessel (ships, yachts) that carries a crew of 15 and more people must be equipped with special holding containers to keep waste (Annex V, MARPOL). During the sailing, solid waste is kept in such containers, and their size is adequate to the number of people on board (converted into m^3). After the sailing, waste is collected from the vessel by port collecting systems in order to recycle or utilize it.

Defined in Annex VI, the regulations refer to prevention of air pollution generated by ships. They refer specifically to the following (Polish Register of Shipping PRS 2022, 41–42):

- operation of onboard reefer equipment that uses ozone-depleting substances
- emission of exhaust fumes that increase air pollution generated by ships
- cargo handling operations involving some types of cargo generating noxious vapors
- incineration of various substances in a shipboard incinerator

Cargo vessels are equipped with refrigerating devices, air-conditioning and fire-protection systems that use substances affecting the ozone layer. The convention regulations forbid the use of chlorofluorocarbons (CFCs), hydro-chlorofluorocarbons (HCFCs) and halons. Installations with refrigerants should be operated in closed circuit to minimize any leaks to natural environment. The convention also requires concentration levels of refrigerating substances to be constantly monitored on vessels. Each vessel should carry a list of equipment and installations that are operated with the use of ozone-depleting substances. There is also an obligation for vessels to keep onboard records of substances that degrade the atmosphere in an Ozone-depleting Substances Record book.

Annex VI provides a possibility to incinerate sludge oil generated during the operation of the main and auxiliary engines and sewage sludge. The document also specifies technical requirements for the operation of shipboard incinerators, where incineration temperature must be constantly monitored as waste should be incinerated at the temperature of 850°C, or 1,562°F (The Marine Environment Protection Committee 2014b).

Safe exploitation of vessels involves numerous technical operations in which ballast systems are particularly significant. A ballast system is crucial for navigational safety of a vessel. The common character of ballasting operations negatively affects marine environment because ships contribute to mixing sea water from various geographical regions. In this way some sea areas may become polluted with strange, non-native flora and fauna. Therefore, adequate regulations were developed to limit pollution of water with non-native species. In 2004 the International Convention for the Control and Management of Ships' Ballast Water and Sediments (BWM Convention) was adopted to provide biological balance (International Maritime Organization 2004). Reduction of biological pollution should be achieved through the use of technical systems for ballast water filtering. The BWM Convention requires each vessel to have a Ballast Water Management Plan for regulating all ballasting operations (Rule B-1). On each vessel an officer must be appointed to be responsible for the implementation of such a plan and for precise definition of safety procedures for proper ballast water management. Another obligation imposed on each vessel is to keep a Ballast Water Record book, where all ballasting operations are logged. The Ballast Water Record book may be inspected by officials appointed by the state administration of each convention member country. A copy of the document can be used as a proof for operations that have been performed on the vessel. Furthermore, the Convention defines the principles for exchanging ballast water, unscheduled discharge, systems of ballast water treatment (adjusting ballast water to the standards set by the convention).

The implementation of the BWM Convention allows interested parties to reduce biological pollution of the sea and to stop the expansion of nonnative species that are carried in ballast tanks. This is a highly significant element to maintain the global balance of marine ecosystems. Such activities comply with the legislation provided in the United Nations Convention on the Law of the Sea—UNCLOS and United Nations Conference on Environment and Development—UNCED.

An important document that complements protection of marine environment in maritime transport is the International Convention for the Safety of Life at Sea—SOLAS. The Convention provides comprehensive regulations on technical and organizational requirements concerning operation and construction of seagoing vessels. Some of these regulations refer to procedures for shipping particular groups of cargo by sea, including the following:

- International Maritime Dangerous Goods (IMDG)
- International Maritime Solid Bulk Cargo (IMSBC)
- Code for the Construction and Equipment of Ships Carrying Liquefied Gases in Bulk (IGC Code)
- Code of Safety for Ships using Gases or other Low-flashpoint Fuels (IGF Code)

All the technical norms and procedures that have been developed to handle particular groups of cargo and types of vessels allow involved parties to maintain a high quality standard of shipping services, with the consideration of natural environment welfare. The implementation of the norms defined by the MARPOL, SOLAS and BWM Conventions provides operation of maritime transport in accordance with the concept of sustainable development (Fitzmaurice 2016, 55–63; Harrison 2016, 181–188). Additionally, the amendments to Annex VI of the MARPOL Convention adopted in 2011 implemented obligatory requirements in the field of energy efficiency of vessels, in order to decrease GHG emission generated by the international maritime shipping during the subsequent years (The Marine Environment Protection Committee 2011). A new Chapter 4 was added to Annex VI of MARPOL Convention. It included legal solutions concerning technical and operational means of energy efficiency of vessels. The solutions were implemented on 1st January 2013, and they are now applicable to all vessels (including container vessels) of the gross capacity above 400 GT in international shipping. The mandatory measures of energy efficiency of vessels are as follows (Pyć 2019, 109–117):

- Energy Efficiency Design Index (EEDI) required for newly constructed vessels and vessels that have undergone major conversions (understood as a technical measure)
- Ship Energy Efficiency Management Plan (SEEMP) required for all currently operated vessels (understood as an operational measure)[3]

The mentioned measures were the first legally binding instruments since the adoption of the Kioto Protocol by the UNFCCC, which referred to GHG emission. They were also the first mandatory global measures for reducing GHG emission generated by ships. Rule 20 to Annex VI of the MARPOL Convention imposes a requirement stating that the achieved EEDI should be calculated for each vessel, in accordance with the guidelines specified by the IMO. The achieved EEDI should be verified on the basis of the EEDI technical files by the flag state or an organization authorized by the flag state (a recognized organization—RO) (Pyć 2019, 106–117). EEDI is not an obligatory solution in terms of technology as the entities obligated to follow EEDI may select technology they wish to apply in the project design of the vessel, providing that the required level of energy efficiency will be achieved and application of economically justified solutions will be possible (International Maritime Organization 2011).

A ship energy efficiency management plan (SEEMP) sets a requirement that entities operating seagoing vessels are obligated to improve energy efficiency of their vessels during their operation through development of a policy for energy-efficiency management. The guidelines provided by the IMO in 2016 on the development of a ship energy-efficiency management plan come as instruments facilitating the development of such a plan in practice (International Maritime Organization 1993). A ship energy-efficiency management plan consists of two parts:

- Part I—presents a possible approach toward the monitoring of efficient operation of a vessel and the entire fleet of a shipowner in time, and it also presents solutions which should be taken into consideration while looking for methods to optimize operation of a vessel.
- Part II—provides methodology which should be applied to collect data in accordance with the Rule 22A of Annex VI of the MARPOL Convention

by vessels of the gross capacity of at least 5000 GT. It also defines the procedures which should be applied by a vessel to report data to the flag state administration (FSA) or to the recognized organization (RO).

The current regulations state that after the implementation of a SEEMP to the Safety Management System (SMS) of a particular shipowner, the monitoring, self-evaluation and improvement of activities should become a part of an audit of the company and of a survey cycles under the ISM Code (Pyć 2012, 243). The implementation of mandatory ship energy-efficiency management plans should be used for regulation of numerous operational problems that are related to shipping undertaken by a seagoing vessel and for which the internationally customary regulations governing the management and control of a ship would not be effective (International Maritime Organization 2011). The compliance of those operations undertaken by a shipowner in accordance with Rule 22 of Annex VI to the MARPOL Convention is to be confirmed by an International Energy Efficiency Certificate.[4]

The IMO has been continuously working on legislation for limiting GHG emission generated by ships since 1997. In 2018, during the 72nd MEPC session, the IMO adopted the *Initial IMO Strategy on reduction of GHG emissions from ships.* In accordance with the strategy, it is planned to reduce GHG emission generated in international shipping by 50% by 2050 in comparison to 2008, with a simultaneous effort aimed at total elimination of GHG emission. As stated in the strategy, the measures for limiting GHG emission generated by ships refer to the following: improvement in energy efficiency, improvement in operational energy efficiency, implementation of market instruments and alternative fuels of low and zero carbon content, such as LNG, hydrogen and similar fuels (The Marine Environment Protection Committee 2018b).

Still, reducing GHG emission generated in international shipping must also refer to activities aimed at reducing emission at seaports (and terminals, including container terminals). In their terminal structure, seaports perform the role of economic and transport hubs, and they are more and more often involved in activities undertaken in favor of reducing GHG emission and protection of natural environment by, for example, participation in the World Ports Sustainability Programme—WPSP (International Association of Ports and Harbors 2022). The *Environmental Ship Index* (ESI) is a project under the WPSP. This index identifies vessels that achieve better results than required by the current emission standards defined by the IMO in the field of reducing emission into the air. The ESI is used for evaluating the amounts of nitrogen monoxide (NOx) and sulphur monoxide (Sox) emitted by a vessel. The ESI also includes reporting GHG emission generated by a vessel. At present, applying the ESI is voluntary. It is a flexible instrument that can be applied to supplement the solutions adopted and developed by the IMO (The Environmental Ship Index (ESI) 2022).

Moreover, the enhancement of a SEEMP has been proposed by the IMO, along with the implementation of a new measure, namely a company energy efficiency management plan (CEEMP) for shipping companies (shipowners). Most shipowners and port authorities have already implemented their Environment Management Systems (EMSs), in accordance with the ISO 14001 standard that includes procedures for selecting optimal measures to reduce emission, methodology for their calculation and control pragmatics. Hence, the monitoring of operational energy efficiency should be treated as an integral element of more expanded systems for management of a shipping company and a port (terminal). This is also related to the fact that there is a relatively large group of factors/components of GHG emission generated by sea shipping that should be monitored and verified (ISO 14064–1 2006).

12.1.3 Components of GHG/CO_2 Emission and CO_2 Equivalent in Maritime Transport (Identification of Factors)

The operation of vessels propelled by internal-combustion engines is a major source of air pollution. Exhaust fumes contain substances that are detrimental for natural environment, such as nitrogen oxides (NO_x), sulphur oxides (SO_x) and particulate matter (PM). Internal-combustion engines that are applied in propulsion systems of vessels must be certified in terms of NO_x emission (Table 12.1).

Reduction of NO_x emission from ship engines has resulted in the search of new types of fuel that can meet the requirements defined in the MARPOL Convention. A gradual increase in strictness of emission standards has brought a necessity of introducing new types of fuel and technical modification of current engines. The solutions that have already been in practical operation include dual-fuel engines, gas-fueled propulsion systems that maintain NO_x emission at Tier II for oil fuel and Tier III for LNG. However, the basic types of marine fuel include hydrocarbon fuel of the parameters presented in Table 12.2.

Emitted during operation of ship internal-combustion engines, sulfur oxide is another detrimental substance for natural environment. In 2008, the IMO amended Annex VI to the MARPOL Convention and tightened the global limits for SOx emission down to 0.50%m/m (the weight share stated in percentage). Considering ecological conditions of some geographical regions, SOx emission control areas were defined and the emission level was decreased to 0.10%m/m (European Parliament and the Council 2016). Sulphur Emission Control Areas (SECA) include: the Baltic Sea, the North Sea, the English Channel, the coasts of the United States of America

---

TABLE 12.1
Total Weighted NO_x Emission from Ship Engines

Requirements concerning reduction of NO_x emission from shingines	Emission parameters
I. Emission tier engines installed on vessels constructed before 1st January 2011	Up to 130 rpm—17.0 g/kWh, 2,000 rpm or more—9.8 g/kWh
II. Emission tier engines installed on vessels constructed after 1st January 2011	Up to 130 rpm—14.4 g/kWh 2,000 rpm or more—7.7 g/kWh
III. Emission tier engines installed on vessels constructed after 1 January 2016	Up to 130 rpm—3.4 g/kWh 2,000 rpm or more—2.0 g/kWh

Source: Data from Polish Register of Shipping PRS (2022, 49–50).

TABLE 12.2

Types and Parameters of Emission for the Specified Types of Marine Fuel

Type of fuel	Emission factors (tCO_2/t of fuel)
Heavy Fuel Oil (HFO)	3.114
Light Fuel Oil (LFO)	3.151
Diesel/Gas Oil (DGO)	3.206
Liquefied Petroleum Gas/Propane (LPG/P)	3.000
Liquefied Petroleum Gas/Butane (LPG/B)	3.030
Liquefied Natural Gas/Propane (LNG)	2.750
Methanol	1.375
Ethanol	1.813

Source: Data from The Marine Environment Protection Committee (2014a).

and Canada, the Caribbean Sea. Tightened standards for SOx emission cause a necessity of replacing currently used fuel with low-sulfur fuel or of applying technical solutions that can reduce pollution from exhaust fumes. Shipowners who operate in the SECA areas use low-sulfur fuel or they install scrubbers that allow them to use regular marine fuel. Both solutions cause an increase in costs of ship operation and the choice of a solution variant depends on shipowners' decisions. The choice of a particular solution also depends on the age of a vessel, a form of how shipping is organized and volatility of fuel prices. There are also other solutions, such as dual-fuel vessels and experimental vessels with electric propulsion drive (for example, the Yara Birkeland project, already discussed in this monograph), which are to meet the NOx and Sox emission standards.

Volatile organic compounds (VOCs) are also a source of air pollution. They can be found in numerous liquid and solid products. Most often, they are chemicals of various types that are used for industrial production and can be found in objects of everyday use (Khan and Ghoshal 2000, 527). Among other products, VOCs can be components of paints, varnishes, medicines, glues, refrigerants, hydraulic fluids, liquid fuels, disinfectants and plastics. VOCs can be released to the air during handling operations involving containers stuffed with products containing those substances.

Over the last decade, it has been widely recognized that GHG emission generated by ships has a direct impact on human health and conditions of living so this problem has eventually come as a challenge to environmental policy makers. Additionally, emission from vessels and the shipping sector contributes to local and regional acidification and eutrophication (part of anthropopressure activities) and also has some influence on so-called "the radiative forcing" (RF) of climate (Corbett et al. 2007, 12–17). An overall impact of shipping emission on the climate change (as a process) is shown in the Figure 12.6.

As a significant element of operating cost, fuel consumption represents an area of special attention for all ship operators and shipowners (Zatouroff and Luke 2013, 4).

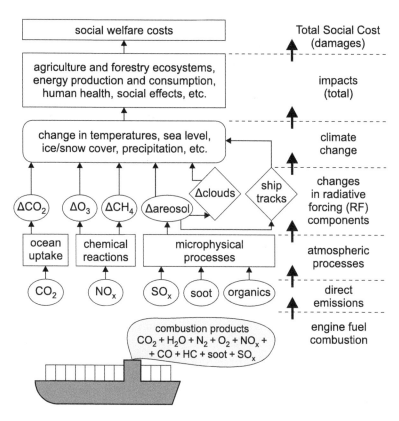

FIGURE 12.6 Schematic diagram of the overall impact of emission from the shipping sector on the climate change.

Source: The authors' own elaboration based on Second IMO GHG Study (2009, 113).

In recent years, through technical and design-based factors, shipping has achieved a noteworthy reduction in fuel consumption, resulting in lower CO_2 emission on a capacity basis (tonne-mile) (Miler and Szczepaniak 2014, 125–126). Additional reduction could be obtained through operational measures, such as lower speed, voyage optimization, etc. (EUROACTIV 2022). However, further reduction of GHG emission generated by maritime shipping would be impossible without implementation of principles and standards for control and verification of emission generated by sea-going vessels (MRV CO_2).

12.2 STANDARDS, CONTROL AND VERIFICATION OF EMISSION GENERATED BY MARITIME TRANSPORT (MRV CO_2/CO_2E)

12.2.1 MRV CO_2—IDENTIFICATION OF THE PROCEDURES, CARBON INTENSITY INDICATOR (CII, EEXI, EEDI)

In June 2013, the European Commission (EC) put forward a legislative proposal to establish a system of monitoring, reporting and verification of CO_2 emission from

Year	2017					2018													2019										
Month	A	S	O	N	D	J	F	M	A	M	J	J	A	O	S	O	N	D	J	F	M	A	M	J	J	A	O	N	D
Actions	1				2	3												4				5	6						
Shipowner	P					Reporting monitoring (process)																							
Verifier DNVGL		Verification																	Verification										
European Commission																						Pub							
MRV process 2018						MRV 2018																							
MRV process 2019																			MRV 2019										

FIGURE 12.7   The MRV Timeline.

Source: Based on European Parliament and the Council (2015a).

Legend:
1—Submission of monitoring plan for verification (31 August)
2—Verified monitoring plan (31 December)
3—Start of first reporting period (1 January)
4—End of first reporting period (31 December)
5—Verified emission report (30 April)
6—Publication of data by EC (30 June)
P—Preparation
Pub—Publication

large ships (MRV) operating within the EU seas and entering EU ports (European Parliament and the Council 2015b). For the first time ever, such as tool (MRV) was designed to build a monitoring system for global shipping emission in order to provide an EU-wide legal framework for collection, verification and publication (on annual basis) of data on CO_2 emission from all ships over 5,000 GT that call at EU ports, irrespective of a place where they are registered (Official Journal of the European Union 2009).

The MRV is the first step in a wider strategy plan, where the second phase includes specifying a reduction target for GHG emission from ships (450 ppm) and finally (as a third phase), an introduction of market-based instruments to the environmental policy.[5] The timeline with the sequence of steps necessary to introduce all the MRV regulations is depicted in Figure 12.7.

This regulation introduces shipowners' obligation of the continuous monitoring of carbon dioxide (CO_2) emission in each cruise ship, starting on 1st August 2018. Based on the monitoring plan assessed in accordance with Article 13(1), shipping companies should monitor CO_2 emission for each ship on a pre-voyage and annual basis, by applying the appropriate method for determining CO_2 emission. In accordance with Article 11, starting from 2019, by 30th April of each year, companies shall submit their emission reports on carbon dioxide emission and other relevant information for the entire reporting period, for each ship under their responsibility, to the EC and to the authorities of the flag states concerned. An annual report should be submitted by each company for each ship for verification performed by a contracted verifier[6] who will assess the conformity with the requirements laid down in Articles 8 to 12 and Annexes I and II to the regulation EU 2015/757 of the European Parliament and the Council. Once the assessment concludes that the emission report is free from material misstatements, the verifier will issue a verification report, stating that the emission report has been verified as satisfactory.[7] An annual report based on the approved monitoring plan assessed by the verifier, should include—but is not limited to—the following information (an example of a report is presented in Annex 1) (Miler and Szczepaniak 2014, 125–126; Akoel and Miler 2019):

- amount and emission factor for each type of fuel consumed in total
- total aggregated CO_2 emitted within the scope of this regulation
- aggregated CO_2 emission from all voyages between ports under a member state's jurisdiction
- aggregated CO_2 emission from all voyages which departed from ports under a member state's jurisdiction
- aggregated CO_2 emission from all voyages to ports under a member state's jurisdiction
- CO_2 emission which occurred within ports under a member state's jurisdiction at berth
- total distance traveled
- total time spent at sea
- total transport work (in tonne-mile)
- average energy efficiency

The IMO has also adopted a three-stage approach to consider further measures for improving the energy efficiency of vessels. The first step is the data collection, then there is the data analysis and the third step is the final decision-making on further measures to be taken by the IMO to implement an appropriate amendment to the MARPOL Convention. The new IMO regulations were adopted during the 70th MEPC session in October 2016, as an amendment to the MARPOL Annex VI, and a new Regulation (22A) was added to the collecting and reporting of ship fuel consumption data (SFCD).[8]

Similary to the MRV, the SFCD data collection system is compulsory for all ships of 5,000 GT and above; ships are required to submit consumption data on each fuel type used on board, with all the additional data specified, including proxies for transport work. The methodology shall be included in the SEEMP plan approved by the flag administration (Akoel and Miler 2019).

The aggregated data should be reported by ship operators to the flag state administration (FSA) after the end of each calendar year. Thereafter, the flag state will submit the collected data to the IMO data centre and, as per the requirements, it will issue a Statement of Compliance to the ship in question. After that, according to Resolution MEPC.292(71) the IMO will produce an annual report to the MEPC, summarizing the collected and anonymised data. The summation of the differences between both adopted systems, namely MRV implemented by the EU Commission and the ship fuel oil consumption data collection (SFCD) implemented by the IMO, is presented in the following table (Table 12.3).

Based on Ship Fuel Oil Consumption Database (The Marine Environment Protection Committee 2016a), calculation and verification of GHG and CO_2 emission generated by maritime shipping and the previously mentioned tools along with ambitious legislative objectives result in growing pressure exerted on shipping companies and shipowners (shipping operators) that is aimed at the application of measures reducing emission (and ultimately keeping it at the minimal level or at the total elimination of emission).

TABLE 12.3

Differences between Two Systems: The MRV Implemented by the EU Commission and the Ship Fuel Oil Consumption Data Collection (SFCD) Implemented by the IMO

Description of the main features	EU MRV	IMO SFCD
Compulsory	Ships 5,000 GT >	Ships 5,000 GT >
	Voyages to/from EU ports of calls	All voyages
	EU Monitoring Plan	Updated SEEMP
	Starting 1st January 2018	Starting 1st January 2019
First monitoring period	2018	2019
Exemptions	Warships, naval auxiliaries, fish-catching/ processing ships, ships not propelled by mechanical means and government ships used for noncommercial purposes	Not defined yet
Parameters to report	Fuel consumption and CO_2	Fuel consumption and CO_2
	Total cargo on board	Deadweight of a ship
	Distance traveled	Distance traveled over ground (O/G)
	Time at sea and in port	Time spent underway
Verification	Independent approved Verifier	Flag administration
Reporting to	European Commission	Flag State Administration
Ceritification	Certificate of Compliance CoC issued by the Verfier, valid for 18 months	Statement of Compliance issued by the flag state, valid for 12 months
Publication	Public database	Anonymous public database

Source: Based on Resolution MEPC.292(71) (2017), Akoel and Miler (2019).

12.2.2 ORGANIZATIONAL, TECHNICAL AND OPERATIONAL FACTORS DETERMINING LOWER EMISSION GENERATED BY MARITIME TRANSPORT

In order to implement the IMO and EU directives, numerous shipowners implement comprehensive ecological programs on their ships, under the SEEMP (including its telematics equipment) to reduce GHG emission generated by exhaust fumes. In such programs, two main groups of factors are taken into consideration:

- Optimization of a sailing (optimal speed of the vessel that means a compromise between the cargo carrier's expectations and the amount of the consumed fuel—see Figure 12.8, the selection of a route with more favorable weather conditions, efficiency of the autopilot and the steering gear, application of the proper trim, that is the difference between the draught of the bow of the vessel and the draught of her stern, the use of ECDIS and Integrated Navigation System/Integrated Bridge INS/IBS, etc.).
- Application of nonstandard technical solutions (covering the hull with layers of silicone anti-fouling paint, polishing the propeller—anti-cavitation, applying special additional substances to fuel, etc.).

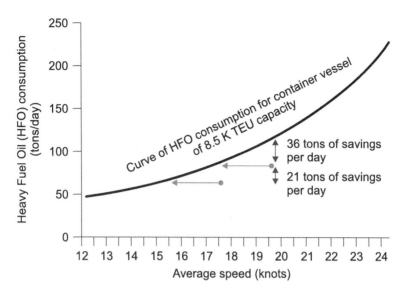

FIGURE 12.8 Dependence of fuel consumption on the speed of a container vessel (8,5 K TEU in tonnes per day).

Source: Based on Igliński (2012, 459).

TABLE 12.4
Potential Reduction of CO$_2$ Emission from Shipping by Using Commonly Applied Technology and Practices

DESIGN (New ships)	Saving (%) of CO$_2$/tonne-mile	Combined design and operation
Concept, speed and capability	2–50[†]	
Hull and superstructure	2–20	
Power and propulsion systems	5–15	10–15%[†]
Low-carbon fuels	5–15*	
Renewable energy	1–10	
Exhaust gas CO$_2$ reduction	0	25–75%[†]
OPERATION (All ships)		
Fleet management, logistics and incentives	5–50[†]	
Voyage optimization	1–10	10–50%[†]
Energy management	1–10	

Source: The authors' own elaboration based on Second IMO GHG Study (2009, 3), Akoel and Miler (2019).

Note: * CO$_2$ equivalent based on the use of LNG; [†] Reductions at this level would require reductions of speed.

Table 12.4 presents an assessment of potential reduction of CO$_2$ emission from shipping by using commonly applied technology and practices.

Implementing the directives of the European Commission on reduction of GHG emission, most vessels have been also equipped with special systems for fuel

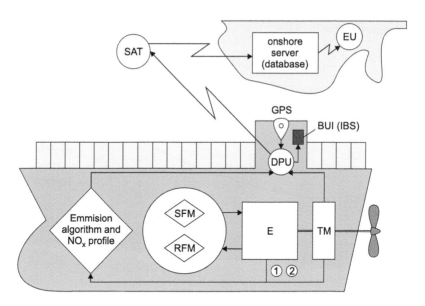

FIGURE 12.9   Onboard system for fuel consumption management and GHG emission (with current monitoring).

Source: The authors' own elaboration based on Safety4Sea (2022).

Legend:

SAT—Satellite Receiver
EU—End User
GPS—Global Positioning System (sensor)
BUI (IBS)—Bridge User Interface (as a part of Integrated Bridge System IBS)
DPU—Data Processing Unit
SFM—Supply Flow Meter
RFM—Return Flow Meter
E—Emission Algorithm and NOx profile (methodology applied)
TM—Torsion Meter
(1)—Engine and fuel specification
(2)—Environmental conditions (from ship's meteo systems)

consumption monitoring that are related to propulsion and power generation (main and auxiliary engines, boilers). The data obtained from those devices are sent to the bridge and to the engine room of a vessel. In modern telematics systems (condition-based monitoring, CBM), the data can be also available to the shipowner's relevant units in real time. In this way the shipowner may decide about selecting the best option in terms of fuel consumption and GHG emission (Miler and Bujak 2014, 57) as well. Figure 12.9 presents an example of a system dedicated to fuel consumption management, connected with a system for monitoring GHG emission (CO_2, NO_2).

A solution that can be applied to streamline processes of fuel consumption monitoring is the APISS Economizer system. The system allows interested parties to measure light fuel oil (LFO) and heavy fuel oil (HFO) consumption in real time and it automatically records that data for further analysis in any spreadsheet format (APPIS 2022). Additionally, owing to the measurement from the receivers installed on the vessel, there is a possibility to register a number of other parameters that are required to balance fuel consumption. The data can be sent to the shipowner by the internet. The system displays the measuring and calculated data, such as the following (Miler and Bujak 2014, 25):

- instant fuel consumption per hour or per nautical mile
- global consumption
- pointer meters for fuel consumption, a course of a vessel, wind direction
- indicators for operating vessel devices, such as boilers, generators, engines
- GHG emission indicator

While selecting which ships to deploy on a particular trade route, ship operators consider costs, time and capacity (IFEU Heidelberg 2011, 47–48). Their objective is to find the optimal mix of the key economic drivers—the least costly route, the shortest distance and the maximum amount of goods that can be transported at any given time.[9] One of the key factors determining this calculation is an environmental factor (GHG emission with obligatory CO_2 monitoring and verification). Figure 12.10

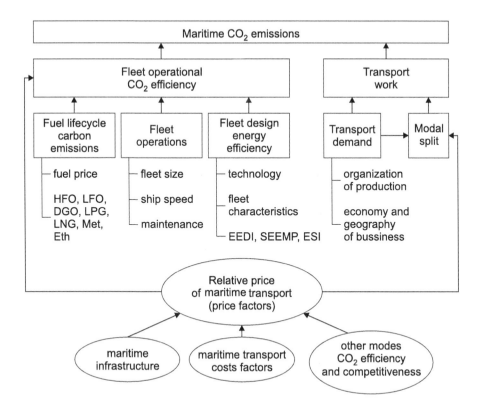

FIGURE 12.10 Factors determining maritime emission of CO_2.

Source: Based on Baltic Maritime Outlook (2006, 114).

indicates all possible factors that can have a potential influence on maritime CO_2 emission.

The future for the shipping industry will undoubtedly involve implementation of the entire spectrum of Green Shipping Practices (GSP) tools (Brzozowska and Miler 2017, 195–209) into its business practices—particularly, in management of the volume and shape of capacity and in requirement for more sophisticated planning and decision-making (involving emission control).

Another anthropopressure factor in maritime transport (next to shipping itself) are seaports, container terminals and their operation that generates CO_2 emission.

12.3 STANDARDS, CONTROL AND VERIFICATION OF EMISSIONS GENERATED BY MARITIME CONTAINER TERMINALS

12.3.1 CARBON FOOTPRINT—IDENTIFICATION OF PROCEDURES AT THE LEVEL OF A MARITIME CONTAINER TERMINAL

After the implementation of legal regulations along with voluntary actions undertaken under corporate social responsibility (CSR), more and more organizations (including maritime container terminals) report and declare reduction of their carbon footprint (CF) (Kulczycka and Wernicka 2015, 61–72). Carbon footprint is calculated as the total amount of CO_2 and other GHG (as CO_{2e}) (EU Ports European Economic Interest Group 2017, 13) in reference to emission resulting from the life cycle assessment (LCA) of a product (service or process, including a shipping process), including its storage and utilization. Carbon footprint can be also calculated with the use of methodologies accepted by companies (including maritime container terminals as well) (WPCI 2009). In both cases, the procedure is the same; however, different elements are taken into consideration in such calculations (depending on the processes involved) (Kulczycka and Wernicka 2015, 61–72). In accordance with the formal guidelines (EU Ports European Economic Interest Group 2017, 13), GHG emission can be calculated with the consideration of three scopes:

- Scope 1—GHG emission from sources of GHG that belong to the organization or its subsidiaries (direct emission)
- Scope 2—GHG emission generated during the processes of power, heat or steam generation that are used by the organization (indirect energy GHG emission)
- Scope 3—GHG emission other than indirect energy GHG emission, which results from the operation of the organization but is generated by the GHG sources that belong or are controlled by other organizations

It means that in the calculation the organization's own direct emission is taken into consideration along with emission generated in the supply chain (Freight Transport Association 2011). Hence, for calculating emission of a maritime container terminal, it is necessary to know the entire operational process of container ship handling at a terminal and the overall functioning of that terminal (logistic processes auxiliary to the main operational processes). Calculation can be provided to the following scopes:

- From cradle to grave—the entire process approached in a holistic way is considered, all the stages, from extraction of raw materials to utilization.
- From cradle to gate—it covers stages from extraction of raw materials to a delivery of readymade products/final service to the customer (including the shipping process).

A more popular method of CF calculation for maritime container terminals is the *from cradle to gate* method. It is much more precise and involves less risk of making mistakes. It allows interested parties to analyze all the processes that can be actually identified and verified by experts. Generally, reporting should take place in accordance with the Global Reporting Initiative (GRI), which assumes presenting the broadest process context along with methodologies for calculating emission and ultimately—carbon footprint (CF).

Considering the results from already existing supply chain research (Garrat and Rowlands 2011), the role of CO_2 emission of containerized sea transport (shipping and maritime container terminals) in the context of the volume of approximately 120M TEU per year (referring to the average trip length of 9,000 km) has been already defined. Based on the shipping model developed by MDS Transmodal and Box Trade Intelligence (MDST 2022), this results in the emission of 1,410 kg of CO_2 per TEU (EU Ports European Economic Interest Group 2017, 16). As a specific effect of the supply chain architecture, most containers do not originate from places located near ports, where the sea leg of transportation starts and terminates. Therefore, a land leg of 200km is included in this calculation. The emission of the land leg (by truck and trailer) is calculated at 1.0 kg per km. This means that the transport of 20ft container (TEU) loaded (an average) with 14.5 metric tonnes of consumer goods results in the emission of approximately 1,850 kg of CO_2, from which approximately 45kg can be attributed to terminal handling—that is just over 2% (EU Ports European Economic Interest Group 2017, 18). Table 12.5 indicates some of the main types of activities within a maritime container terminal in order to demonstrate a range of emission and scopes it falls within.

The vast majority of emission reported by maritime container terminal operators is of Scope 1 and 2. At present, most operators do not monitor Scope 3 emission; however, some companies have just started to measure certain elements of Scope 3 emission, for example, air and rail business trips of their staff (not officially reported yet).

Figure 12.11 presents an exemplification of a maritime container terminal operational layout with color identification of procedures and operations included in GHG footprint calculation.

TABLE 12.5

Examples of Activities under the Scopes (1–3) of GHG Footprint (Carbon Footprint CF)

Scope	Activity causing emission	Source of emission
SCOPE 1 (direct emission)	RTG moving containers at the maritime terminal (impact within the terminal boundary)	Diesel powered engine
	RS handling containers at the maritime terminal (impact within the terminal boundary)	Diesel powered engine
	SC handling containers at the yard (impact within the yard boundary)	Diesel powered engine
	ECH (Empty Container Handler) transporting empty containers within dedicated positions (impact within the terminal boundary)	Diesel powered engine
	MHC (Mobile Harbor Crane) as a supporting suprastructure handling containers (impact within the terminal boundary)	Diesel powered engine
	TT handling containers at the yard (impact within the yard boundary)	Diesel powered engine
	AGV handling containers at the yard (impact within the yard boundary)	Diesel powered engine
	Heating Units (buildings, workshops, warehouses etc.) (impact within the terminal boundary)	Gasoil powered heating units causing combustion
SCOPE 2 (indirect emission)	QC (STS) handling containers from and into the container vessel (impact at the power station, off-site)	Electrically-powered engine, however supplied by an external utility company (not always green energy)
	RMG moving containers at the maritime terminal (impact at the power station, off-site)	Electricity used to power RMG
	ASC (Automated Staking Cranes) (impact at the power station, off-site)	Electricity used to power ASC
	AGV handling containers at the yard (impact at the power station, off-site)	Electricity used to power AGV
	Recharging electric forklifts at the Container Freight Stations (CFS) (impact at the power station, off-site)	Electricity used for battery charging
	Terminal security and operational lighting (impact at the power station, off-site)	Electricity used to power lights
	Air condition devices in the offices (impact at the power station, off-site)	Electricity used from the grid
	Other suprastructural equipment (terminal vehicles, SCs, MHCs etc.) of the electrical power (impact at the power station, off-site)	Electricity from the grid/ storage batteries (stations)
SCOPE 3 (optional indirect emission)	Staff business trips, e.g., traveling by different means of transportation from the terminal to the offices at a different location (impact at the transport modes)	Mileage and impact of transport modes
	Staff commuting (usually by owned cars, buses, trains etc.) (impact at the transport modes)	Mileage and impact of transport modes
	Stationery used in offices (Life Cycle Assessment LCA) (impact at the production and utilization)	Production and use of certain products and its contribution to emission as a CF

Source: EU Ports European Economic Interest Group (2017, 23).

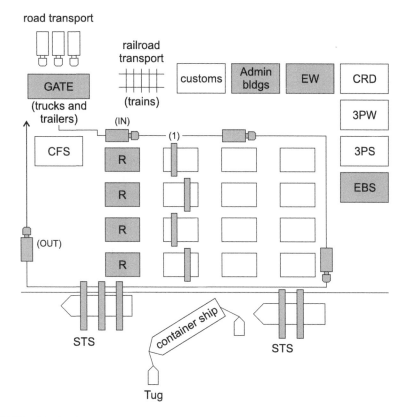

FIGURE 12.11 Scheme of maritime container terminal operations included into CF calculation.

Source: Based on EU Ports European Economic Interest Group (2017, 24).

Legend:

Gray boxes—emission included
White boxes—emission excluded
(1)—all suprastructure equipment eg. RTG, RS, SC, SCH, etc.
CFS—Container Freight Station
Admin Bldgs.—all office and admin buildings at terminal
EW—Engineering Workshops
CRD—Container Repairing Depot
3PW—Third Party Workshop
3PS—Third Party Storage
EBS—Empty Boxes Storage
R—Reefer containers

According to the GHG Protocol (WRI/WBCSD 2006) recommendations, a maritime container terminal should report the following (EU Ports European Economic Interest Group 2017, 24):

- total Scope 1 and 2 emission
- emission data separately for each Scope

- emission data for all seven GHGs separately (CO_2, CH_4, N_2O, HFCs, PFCs, NF_3, SF_6) in tonnes of CO_2 equivalent (CO_2e)
- year chosen as the base year
- an appropriate context for any significant emission changes that trigger the base year emission recalculation (acquisitions/divestitures, outsourcing/insourcing, changes in reporting boundaries or calculation methodologies, etc.)
- emission data for direct CO_2 emission from biologically sequestered carbon (e.g., CO_2 from burning biomass/biofuels), reported separately from the Scopes
- methodologies used for calculation or measurement of emission, providing a reference or a link to any calculation tools applied
- any specific exclusions of sources, facilities and operations

The scope and meters, along with methodologies for calculating carbon footprint at DCT Gdańsk, Poland (a member of PSA Group), are presented in Table 12.6.

TABLE 12.6

The Fields of the Holistic Analysis of Factors Generating CO_2 Emission in the Operation of DCT Gdańsk (PSA Group)

Goal	Indicator	Description/methodology
Terminal potential	ABU (Average berth utilization)	Calculation: % of ((Vessel length + Margins) x Sum hrs of terminal)/Sum hrs in week
Terminal potential	AYU (Average yard utilization)	Calculation: % of Used slots/Total slots
The quality of gate complex service	TTT Truck turn time [min]	Calculation: Sum Truck transactions time/ Sum Trucks
Good place to work	Staff turnover (employee initiated)	N/A
Work safety	LTIFR (Lost Time Injury Frequency Rate)	(Number of lost time injuries in the reporting year x 1,000,000)/Total hours worked in the reporting year
Safe workplace	ASR (Accident Severity Rate)	Number of man-days lost due to workplace accidents per million exposure hours
Environmental friendliness	CO_2 emission [kg/Physical TEU (YTD)]	
CO_2 Emission	KG CO_2	Scope 1 and Scope 2 emission
Water consumption	WU (Water Usage)	Water (liters)/Worked Hours
QC failure	MBBF QC (Mean Box Between Failure)	elapsed crane moves between inherent failures of a QC (STS), during normal operation
RTG failure	MMBF RTG (Mean Moves Between Failure)	elapsed crane moves between inherent failures of a RTG cranes, during normal operation
RMG failure	MMBF RMG (Mean Moves Between Failure)	elapsed crane moves between inherent failures of a RMG, during normal operation

Source: Data from DCT (2022), Dominika million, head of Sustainable Development Department, DCT.

Obtaining data from operational fields of terminals allows interested parties to provide stricter emission monitoring and, as a result, to apply a number of (organizational, technical and operational) solutions that determine their capability to lower emission and carbon footprint in the field of operations undertaken by maritime container terminals.

12.3.2 ORGANIZATIONAL, TECHNICAL AND OPERATIONAL FACTORS DETERMINING LOWER EMISSION GENERATED IN THE OPERATIONAL FIELD OF A MARITIME CONTAINER TERMINAL (SCOPE 1 AND 2—SCOPE 3 ULTIMATELY)

Most maritime container terminals have already developed their strategies (or at least draft strategies) to achieve zero-emission by 2050 (in accordance with the objectives of the *Fit for 55* concept). For example, during the implementation of its strategy, the PSA (the majority owner of—among others—DCT Gdańsk, Poland) has already implemented a Climate Response Management System (CRMS) to the organizational structures and operational processes at all its terminals.

The PSA Group implements solutions in the field of STS electrification and other suprastructural facilities (including RTGs, TTs) and resigns from the current drive systems based on engines powered by diesel fuel. There have been some experiments carried out with alternative types of hydrogen fuel as well.

The PSA Singapore has entered into joint initiatives with multiple government agencies and corporations to study and pioneer ways to utilize hydrogen as a viable low-carbon energy source. The PSA Antwerp has begun trials with a hydrogen-powered TT and a mobile hydrogen refilling station and the PSA Marine has successfully deployed two dual-fuel LNG terminal/harbor tugs (it is expected that it may reduce CO_2 emission by 20%).

The element that lowers emission through the optimization of processes is The Vessel Pilot Communication (VPC)—a PSA ONEHANDSHAKE module responsible for informing vessel captains about the exact time and place of taking a pilot on board. In this way, vessels can maintain adequate speed and enter the port/terminal according to their ETA. It allows vessels to optimize fuel consumption and to reduce CO_2 emission (lowering CF). Another initiative is a recommendation for switching (where possible) to modes of transport characterized by lower emission (a terminal feeder function). Ashcroft Terminal in Canada is a perfect example at this point. It is one of the biggest container terminals in British Columbia that recommends the use of railway transport instead of road transport. Similar recommendations are given by container terminals in Singapore, China, Belgium, India, Italy and Poland. It allows them to implement the UN Sustainable Development Goals (17 SDGs) in a more efficient way, pursuing the objective of lowering emission by 2030.

In accordance with the concept presented by the PSA International Group, the implementation of the strategy at the level of terminals is related to the implementation of solutions that lower emission in the four key areas of action (PSA International Sustainability Report 2020):

1. taking climate action
2. transforming supply chains
3. nurturing a future-ready workforce
4. stewarding responsible business

The processes related to proper steps undertaken in the key areas defined in the previously mentioned way (on the example of the PSA Group) are presented in Table 12.7.

TABLE 12.7

Targets and Commitments of the PSA Group in the Field of Sustainability of Terminals and the Net Zero-Emission Strategy

Key field	Targets/commitments	Actions	UN 17 SDGs contribution
Taking climate action	Emission: • Absolutely reduce Scope 1 and 2 carbon emission by 50% by 2030 and by 75% by 2040 against 2019 as a baseline year, • Achieve net zero emission by 2050, • Establish Scope 3 inventory by the end of 2020 (as a first step toward settlement of Scope 3 emission reduction target. Energy: • Aim at 90% RTGs to be electric (or hybrid) by 2030, • Procure only green powered RTGs from 2023 onward	1/ explore and test the viability of cleaner, renewable energy sources (e.g., wind, solar, hydrogen), 2/ continue converting container handling equipment (suprastructure of terminals) from diesel to electricity powered, 3/test other forms of electrification (electric prime movers, battery powered AGVs), 4/source a greater proportion of electricity needs from green energy sources (e.g., cold ironing)	7, 9, 13
Transforming supply chains	Optimization of global supply chains: • Implement 10 supply chain projects providing sustainable logistics and transport (impacting at least 30K TEU of cargo volume) by 2024, Innovation and technology: • Invest at least USD 100 million by 2025 in R&D to achieve more efficient and sustainable operations.	1/ provide sustainable supply chain solutions to cargo owners, operators and service providers, 2/continue to develop functionalities of the CALISTA digital platform, 3/partner other stakeholders to implement digitalization in the main processes	7, 9

TABLE 12.7

Targets and Commitments of the PSA Group in the Field of Sustainability of Terminals and the Net Zero-Emission Strategy (continued)

Key field	Targets/commitments	Actions	UN 17 SDGs contribution
Nurturing future-ready workforce	People development: • Achieve at least 75% participation rate in the global Employee Opinion Poll (EDP), Occupational Health and Safety: • Aim at zero significant incidents (zero fatalities and permanent disabilities)	1/provide adequate training and personal development opportunities to employees (future-ready skills), 2/promote corporate culture and behavior to meet future needs, 3/continue to enforce health and safety measures	8, 9
Stewarding responsible business	Sustainable port development: • Implement PSA recommendations for sustainable concrete for 50% of new civil infrastructure construction project (over USD 60 million in value) by 2023 and 80% by 2030, Cybersecurity and data privacy: • Adopt cybersecurity best practices, Port security: • Ensure highest possible standards at all terminals (incl. ISPS regulations), Ethics business conduct: • Conduct business with highest standards, • Fraud zero tolerance	1/work closely with relevant authorities in order to embed sustainability into any port/terminal infrastructure development, 2/incorporate new sustainability requirements for construction of new building (e.g., including passive solutions), 3/establish a high level of cybersecurity standards through the implementation of Cybersecurity Management System (CMS), 4/continue to manage port security according to the Safety Security and Environment Management System and international best practices, 5/ensure compliance with the Code of Business Ethics and Conduct, 6/ensure whistleblowing policy is in place	11, 12

Source: Data from PSA International Sustainability Report (2020).

Another challenge is posed by the previously mentioned problems concerning increased emission that results from the operational expansion of the terminals, despite the fact that ecological solutions have been implemented. Hence, the PSA Group indicates the emission level slightly higher in 2020 (732K t CO_2e) than it was in 2019 (728K t CO_2e). The data on the components of the calculated CO_2e emission in the PSA Groups are presented in Table 12.8.

The increase of the total emission by approximately 0.5% resulted from the operational expansion onto the intermodal area and an increase in reefer container handling operations, accompanied by a longer time period required for container handling caused by the Covid-19 pandemic turbulences observed at numerous ports

TABLE 12.8

Total Emission of the PSA International Group 2019–2020 (K t CO$_2$e)

Scope	2019	2020
Direct Scope 1 emission	496	485
Indirect Scope 2 emission	232	247
Total	**728**	**732**

Legend: Scope 1 and 2 emission is computed with use of eqiuty share consolidation methodology, calculations cover CO$_2$, CH4, N20; emission factors for fuel sourced from the
GHG Protocol Emission Factors for cross sector Tools (2017). Additionally, Scope 2
emission is computed with the use of the respective market-based emission factor.

Source: Data from PSA International Sustainability Report (2020, 23).

TABLE 12.9

Total Energy Consumption of the PSA Group 2019–2020 (TJ/MWh)

Energy sources	2019	2020
Fuel from non-renewable sorces (diesel, LNG, petrol, CNG, LPG, biofuels) [TJ]	9,600	9,600
Electricity [MWh]	974,800	1,073,200
Total (equivalent)	13,100	13,500

Source: Data from PSA International Sustainability Report (2020).

(terminals) in the world. The Scope 1 and 3 emission profiles at the container terminals of the PSA Group are as follows (PSA International Sustainability Report 2020):

- Scope 1 (63% of total emission), top 3 emission sources: prime movers (38%
  of Scope 1 emission), diesel yard cranes (21%), marine vessels (19%)
- Scope 2 (37% of total emission), top 3 emission sources: QC quay cranes STS
  (30% of Scope 2 emission), reefer points (30%), electric yard cranes (18%)

The PSA Group recorded a light increase in the field of energy consumption from
13,100 TJ in 2019 to 13,500 TJ in 2020. The data concerning components liable for
energy consumption in the PSA Group are presented in Table 12.9.

In order to reduce the total emission in the subsequent years, in accordance with
the assumed strategy, the PSA Group intends to implement the following solutions
(PSA International Sustainability Report 2020):

- Scope 1 abatement by:
  - low carbon fuel and/or electrification (examples: use of low carbon fuels,
    hybrid and electric power, dual-power tugs, electric STSs, introduction
    of variable speed drive control for diesel-powered RTGs, use of LED
    lights for cranes and buildings, light optimization for automated yard
    cranes, an automatic switching-off system for inactive equipment and

offices, workflow optimization for prime movers to reduce unproductive energy consumption); and

- optimization and energy efficiency (examples: implementation of a programme to reduce waste, starting work on hydrogen as alternative fuel).
- Scope 2 abatement by:
  - purchase of renewable energy (examples: the power purchase agreement (PPA), use of green energy to supply ships while at the terminal—share powering/cold ironing);
  - generation of renewable energy (examples: investment in renewable energy assets, such as (solar) photovoltaic panels);
  - electrical grid optimization (examples: introduction of a smart grid based on artificial intelligence solutions—Smart Grid Management System—SGMS, introduction of a battery energy storage system BESS); and
  - storage and transport of renewable energy (examples: introduction of energy vectors).
- Ultimately, Scope 3 abatement by:
  - emission reduction projects (examples: participation in joint green projects, e.g., Forestry, Carbon Capture and Utilization CCU, Carbon Capture and Storage CCS).

At each PSA Group terminal, units established under the CRMS carry out research and implement solutions in the field of emission reduction. Modern tools from the area of emission control and forecasting paths for emission reduction are broadly applied—e.g., Marginal Abatement Cost Curve (MACC). Emission is monitored and reported in real time by the particular terminals to aggregate the results and to provide their analysis at the level of the entire PSA Group.

The aspect referring to emission reduction of Scope 3 introduces a necessity of closer cooperation under the entire logistics supply chain (a part of which is a maritime container terminal).

12.3.3 A Maritime Container Terminal as an Element of a Low-/Zero-Emission Sea-Land Logistic Chain

A holistic concept that displays the functioning of the container terminals belonging to the PSA Group is presented in Figure 12.12.

Considering a comprehensive approach toward tools that decrease anthropopressure and a pursuit of zero-emission at the container terminals of the PSA Group, two key concepts can be observed in the field of process digitalization (PSA International Sustainability Report 2020):

- Internet of Logistics (IoT-like) as a concept that generates an industrial ecosystem of linked logistics operations (that are characterized by interoperability and full synchronization) implemented by all the stakeholders of the logistics chain (regardless of the mode of transport)
- the use of the CALISTA (Cargo Logistics Inventory Streamlining and Trade Aggregation) platform (as a system for visualization and optimization of flow management and, ultimately, a system that functions as an emission calculator—after the implementation of adequate methodologies)

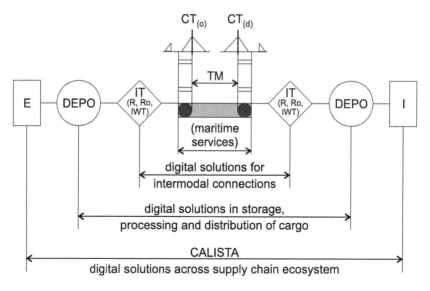

FIGURE 12.12 Operational Value Chain of PSA terminals.

Source: PSA International Sustainability Report (2020, 10).

Legend:

$CT_{(o)}$, $CT_{(d)}$—Maritime Container Terminal: (o) origin, (d) destination
TM—Terminal Management (supported by the use of TOS)
Maritime Services—pilotage, crew transfer, ship-to-ship transfer etc.
$IT_{(R, Ro, IWT)}$—Inland Container Terminal: Road, Rail, Inland waterway
DEPO—Depots (container), freight stations, warehouses etc.
E-Exporter (also: producer, consigner)
I—Importer (also: consumer, consignee)

An example of using the CALISTA platform at the PSA Group terminals is presented in Figure 12.13.

This solution optimizes the flow of documents (EDI) under the CALISTA system, with the particular consideration of intermodal operations (maritime shipping/inland waterway shipping) in the operation of maritime and inland waterway terminals. It allows interested parties to provide e-documents precisely on time in the e-VGM form (Verified Gross Mass) for each container that is being handled at the terminal. Such an approach has already resulted in higher attractiveness of an intermodal connection with the use of inland waterway shipping (instead of the current connection based on road transport). As a result, emission has been reduced (considering the significantly lower carbon footprint of inland waterway shipping).

The problems related to the calculation of Scope 3 emission and a holistic approach toward emission under the entire logistic chain come as relatively new issues introduced quite recently into the pragmatics of the functioning of maritime

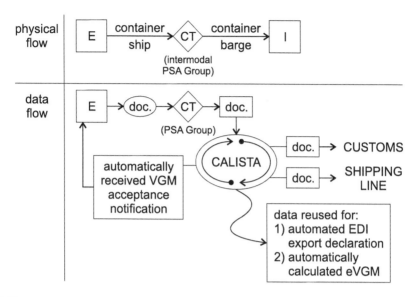

FIGURE 12.13 Achieving an optimized, connected and sustainable supply chain in the PSA Group.

Source: PSA International Sustainability Report (2020, 40).

Legend:

E-Exporter (producer, consignee)
I—Importer (consumer, consigner)
CT- Intermodal Container Terminal (the PSA Group)
(doc)—electronically generated document
EDI—Electronic Data Interchange system
VGM—Verified Gross Mass, eVGM—e document

containerized transport and maritime container terminals. Therefore, it is necessary to implement some advanced work aimed at obtaining the assumed zero-emission outcome by 2030(50). A methodology that can be applied in this process is presented in Figure 12.14.

The considerations referring to the further potential of reducing emission at maritime container terminals and entire logistics supply chains, which come as their parts, must be continued because the aims faced by the discussed sector are extremely demanding. Humanity (politicians, business people, society) has been procrastinating on the implementation of efficient solutions that can prevent negative changes to natural environment. Considering the fact that—according to numerous ecological organizations, there is very little time left, the suggested changes must be radical and the expected results must be clearly visible. Unfortunately for the sector of maritime containerized transport, the activities discussed in this chapter come just as the beginning of a long and difficult way to zero emission.

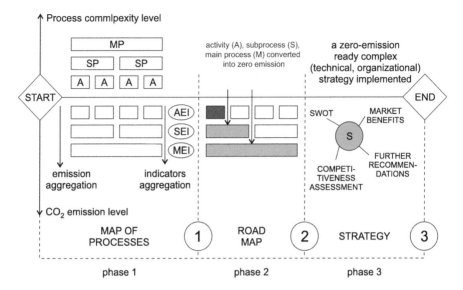

FIGURE 12.14 Stages of the process of achieving zero-emission at maritime container terminals.

Source: Authors' own elaboration.

Legend:

1. End State of Phase 1: a detailed map of processes at a maritime container terminal prepared; each process is supported by a dedicated CO_2 emission calculation methodology.
2. End State of Phase 2: a road map on methods to convert processes (high emission–low emission–zero emission) in order to achieve zero-emission at a maritime container terminal prepared and implemented.
3. End State of Phase 3: a strategy on sustainable development (with zero-emission parameters) prepared and introduced (in order to maintain and sustain zero-emission operations while a maritime container terminal is being developed, its territory expanded and the volume of its activities extended—ready for the future).

AEI, SEI, MEI—Activity, Subprocess, Main Process Emission Indicator

NOTES

1. The session resulted in the publication of 26 postulates concerning the protection of the natural environment and ecological policy. The guidelines included postulates on providing international cooperation in the field of developing policy for the protection of natural environment, implementing systems for monitoring natural environment and cooperating and participating in legislation of the international law for natural environment protection.
2. Total nitrogen: 20 Qi/Qe mg/l or at least 70% reduction; total phosphorus: 1.0 Qi/Qe mg/l or at least 80% reduction.
3. In accordance with Rule 22 of Annex VI to the MARPOL, SEEMP may be a part of the Safety Management System (SMS).
4. In order to facilitate the implementation of mandatory measures of ship energy efficiency, the IMO has adopted numerous guidelines, including guidelines on the method of calculation of the attained energy efficiency design index (EEDI) for new ships (Res.

MEPC.308(73); guidelines for the development of a ship energy efficiency management plan (SEEMP) (Res.MEPC.282(70); guidelines on the survey and certification of the energy efficiency design index (EEDI) 2014 MEPC.261(68) and MEPC.309(73), a consolidated version in a form of a circular MEPC.1/Circ.855/Rev.2(Res.MEPC.254(67); guidelines for the calculation of reference lines for use with the energy efficiency design index (EEDI) Res.MEPC.231(65) (The Marine Environment Protection Committee 2013, 2016b, 2018a, 2019).

5. The implementation of the first of these activities is a regulation of the European Parliament Council (EU) 2015 sig/757 on 29th April 2015 in terms of monitoring, reporting and verifying of CO_2 emissions from maritime transport and amending the provisions of Directive 2009/16/EC sig (MRV)—Official Journal of the European Union L 123/57.

6. Verifiers should be independent, and competent legal entities and should be accredited by the national accreditation bodies established under Regulation (EC) No 765/2008 of the European Parliament and the Council, www.imo.org/MediaCentre/HotTopics/GHG/Pages/default.aspx (accessed: 2nd March 2022).

7. The verifier shall issue, on the basis of the verification report, a document of compliance for the ship concerned, valid for 18 months after the end of the reporting period.

8. The data collection system is enshrined in the amendments to the International Convention for the Prevention of Pollution from Ships (MARPOL) Fuel oil Consumption Database' (Appendix IX), Fuel Oil Consumption Reporting (Appendix X). These amendments entered into force on 1st March 2018, and the first reporting period was for the 2019 calendar year.

9. The type of goods being transported also matters. Raw commodities, which are handled with very thin margins, generally go the least costly routes; high-value, time-sensitive or perishable goods will take the shortest routes.

13 Conclusions

Introducing containers in maritime transport has been a revolutionary change to the current systems applied to ship general and semi-bulk cargo. Changes have been observed in numerous organizational procedures related to cargo shipping and in the implementation of new transport and cargo handling technologies. At first, containerization referred mainly to maritime transport, where general cargo handling processes were streamlined on the port-vessel-port line. Hence, the necessity of arranging numerous dock workers and using auxiliary cargo handling devices was reduced. In the 1970s, containerization started to perform a significant role in general, semi-bulk and, later on, bulk cargo handling in merchandise trade involving all other modes of transport. The impact of maritime container transport resulted in gradual implementation of the container system, most of all through the construction of port terminals adjusted to handle container vessels. As nodal infrastructure facilities, port terminals were equipped with special suprastructure dedicated to handle containers of standardized types. In the system understood in this way, railway and road modes of transport started to provide feeder services to port container terminals.

The popularization of container transport in the world took place owing to technical standards and standardized container dimensions that were implemented by the International Organization for Standardization in 1968. The implementation of the global standards for this type of transport devices allowed interested parties to develop a fleet of container vessels and port container terminals. Owing to dimensional standards compliance, it was possible to create a global system of containerized cargo shipping that can now guarantee the efficient handling of vessels and cargo at various seaports.

The implementation of containers resulted in a dynamic development of the global container vessel fleet. Taking the economies of scale in their consideration, shipowners actively invested in their fleets, which resulted in an increase in the unit loading capacity and deadweight tonnage of vessels. It was also possible to observe a sudden increase in the size of container vessels. The first rapid increase took place in 1997, when container vessels reached an average size of 8,000 TEU. The second one started in 2006, when vessels of the average deadweight tonnage of 15,500 TEU were introduced into operation. Over the last decade, the unit loading capacity of individual container vessels has already reached the level of 24,000 TEU.

The development of maritime container shipping was accompanied by adjustments made at maritime ports to provide adequate depth at port wharves and increased capabilities in the field of cargo handling and storing at port container terminals. The optimization of terminal suprastructure was also of great significance as it was done to achieve high efficiency in the flow of goods in the vessel-terminal and terminal-vessel relations. Nowadays, a highly important factor in the functioning of port terminals is their connection with the hinterland, which allows containers to be efficiently delivered to their destination places.

DOI: 10.1201/9781003330127-18

Involving numerous means and modes of transport into shipping processes allows interested parties to develop a wide-ranged, highly efficient interdependent transport system. Such a solution has come as an inspiration for the development of the container transport system, which is based on transport chains developed between the consigner and the consignee of the cargo to provide a possibility of implementing the just-in-time concept. Its theoretical assumptions have been implemented in the form of multimodal and intermodal transport, where one operator handles the entire shipping process, starting at the place where the consignment is dispatched and ending at the place where it is received. In this case, the necessity of opening containers and deconsolidation of containerized cargo is minimized, which is good for the cargo, and it allows interested parties to reduce cargo handling costs. Containers offer possibilities to organize and to implement shipping processes globally, which allows container shipping chains (supply chains) to be developed regardless of the places where the consigner's and consignee's headquarters are located. Such a solution is beneficial to all the participants of the supply chain because each participant obtains their benefits and the efficiency of the entire process is improved. This mainly refers to the involvement of maritime and land carriers, port container terminals and land terminals (dry ports). While achieving their economic benefits, each link of this chain generates added value, and the entirety of activities can be observed in the form of high quality shipping services. The integration of transport is also fostered by legal solutions, including shipping documentation pertaining to the use of containers in maritime and multimodal/intermodal transport. It is also more often possible to observe a tendency to implement improvements for booking parties or parties that enter into shipping contracts based on the multimodal/intermodal bills of lading.

The container transport system is able to handle shipping processes on a global scale, regardless of shipping directions, geographical regions and seasonal character of consignments. The main container shipping routes are: East Asia–North America, East Asia–Europe and Europe–North America (Review of Maritime transport 2019, 15). It is possible to indicate East Asia as a dominating center of containerized cargo export to other continents. This fact is related to industrial production in Japan, South Korea, Vietnam, Taiwan and the largest manufacturer of consumer goods, China. It seems that in the nearest future the production and consumption centers will remain unchanged, even with the consideration of a lower demand for goods, which is related to the Covid-19 pandemic. In the context of the pandemic, the problem concerning the reconfiguration of global containerized supply chains requires some special consideration. It is necessary to eliminate problems caused by the "key link" (with the lack of any alternative solutions) that brings on potentially serious delays in scheduled flows of raw materials in the supply logistics dendrite. The automotive sector might come as a graphic example at this point, with the reference to its problems with the availability of semiconductors. These problems result in considerable delays in manufacturing new vehicles and even in the necessity of stopping the production at plants all over the world.

The costs of maritime container transport are significant factors that affect the efficiency of all operations, and they are translated into the prices of services provided to customers. The prices of maritime container freight are also affected by economic conditions and all changes that occur to the market. Considering current turbulences related to the pandemic, the costs of maritime freight have soared and reached their highest levels that were rarely observed in the past. Furthermore, the situation is not any better because of a decrease in the availability of empty (and new) containers and problems related to their repositioning.

This comes as another challenge that must be faced and reprogrammed in the modern reconfiguration of supply chains after the post-pandemic analysis, if the modern world of global trade and transport really want to draw conclusions from the global pandemic it has experienced.

An efficient container transport system becomes more and more dependent on the application of modern IT technologies. They are indispensable to provide the flow of information about containers and to implement physical shipping processes and efficient management. Monitoring systems are necessary to handle containers and to indicate the position of individual containers, their cargo and the condition of that cargo (e.g., remote systems of temperature monitoring). Planning loading operations for vessels with the use of IT systems (vessel-planning systems) is also highly important. These systems allow users to link the particular IT processes with physical cargo handling operations. All the operations and operational processes that take place at container terminals are managed by terminal operation systems, which initiate and supervise the course of all the operations involving containers at the facility premises. Each container terminal provides access to its system through the use of special software. Owing to this software, forwarders, consigners and various carriers are able to enter all the required data pertaining to the cargo and to monitor the flow of containers. In the nearest future, such systems are expected to generate further potential in the field of process optimization, also based on ecological optimization (the imperative of lowering anthropopressure through emission monitoring, decarbonization, electrification and an increase in the coefficients of ecological efficiency).

Maritime container transport comes as an important element of the global economy that allows interested parties to implement shipping processes in the global scale. In 2020, the level of maritime container shipping was increased by 6.9% in comparison to the data recorded in 2019, and it should be expected that a decrease related to the crisis caused by the Covid-19 pandemic will be partially made up for in 2023. However, the condition, the potential and the possibilities of further development and, generally, the future of the global container transport system and its powerful link that is maritime transport of containerized cargo depend on identification and profound analysis of factors that have negatively affected the global system of container transport and that have been clearly exposed by the pandemic and military conflicts, such as a war in Ukraine. The previously mentioned elements also depend on the far-reaching, bold conclusions that can be drawn from the current situation and on the implementation of adequate repair solutions.

Appendix

*MRV Annual Emission
Report under EU Regulation
2015/757 as Amended*

Ship	IMO	Built	DWT	GT	Type	MRV Ship Type EU Regulation 2015/757 as amended	Fuel type	Total distance traveled N.miles	Total Fuel Consumption m tonnes	Total CO₂ emissions m tonnes	Time Spent at Sea hours	Time anchorage hours	Total transport work (DWT). N miles	Fuel consumption per distance kg/N mile	Fuel consumption (g) per transport work (dwt)	CO₂ emissions per distance kg/N mile	CO₂ emissions (g) per transport work (DWT) N mile
BBC Amethyst	9504724	2011	14,800	12,838	Heavy Lift Vessel/Tweendecker	General cargo ship	HFO; MGO; MDO	55,756.10	4,274.30	13,361.5936	4,352.710	291.65	505,619,440	76.6607	8.4536	239.6436	26.4262
BBC Citrine	9504748	2012	14,800	12,838	Heavy Lift Vessel/Tweendecker	General cargo ship	HFO; MGO; MDO	18,382.00	No EU call	No EU Call	15,382.000	1,241.90	130,263,732	71.1156	8.3976	221.8278	26.1942
BBC Danube	9571399	2013	18,000	12,974	Heavy Lift Vessel/Tweendecker	General cargo ship	HFO; MGO; MDO	32,109.00	2,070.14	6,542.9734	2,563.140	316.85	322,047 983,7	64.4723	6.428	203.7738	20.3168
BBC Emerald	9504750	2013	14,800	12,838	Heavy Lift Vessel/Tweendecker	General cargo ship	HFO; MGO; MDO	15,382.00	1,093.90	3,412.1546	1,241.900	0.00	130,263,732	71.1156	8.3976	221.827	26.1942
BBC Hudson	9435868	2009	17,500	12,936	Multi Purpose Vessel/Tweendecker	General cargo ship	HFO; MGO; MDO	28,971.00	2,044.03	6,468.866	2,198.950	165.15	416,067 378.22	70.5543	4.9127	223.2876	15.5476
BBC Nile	9571375	2011	18,000	12,974	Heavy Lift Vessel/Tweendecker	General cargo ship	HFO; MGO; MDO	28,750.00	2,062.32	6,514.1141	2,466.590	62.24	351,429 439,58	71.7329	5.8684	226.5779	18.5361
BBC Parana	9571387	2012	18,000	12,974	Heavy Lift Vessel/Tweendecker	General cargo ship	HFO; MGO; MDO	20,352.00	1,393.88	4,405.505	1,580.120	66.05	231,539 118,24	68.4886	6.0201	216.4655	19.027
BBC Russia	9700392	2018	12,500	11,492	Heavy Lift Vessel/Tweendecker	General cargo ship	HFO; MGO; MDO	40,820.00	1,683.91	5,331.7261	3,168.390	521.79	273,197 787,28	41.2521	6.1637	130.6155	19.615

Ship	IMO	Year	GT	Type	DWT	Category	Fuel	Col9	Col10	Col11	Col12	Col13	Col14	Col15	Col16	Col17	Col18
BBC Seine	9508380	2010	18.000	Heavy Lift Vessel/Tweendecker	12.974	General cargo ship	HFO; MGO; MDO	25,171.00	1,718.06	5,429.1313	1,981.740	57.82	243,244 349,3	68.2555	7.0631	215.6899	22.3197
BBC Louise		2017	11.900	Heavy Lift Vessel/Tweendecker	9.700	General cargo ship	HFO; MGO; MDO	15,012.00	765.67	2,418.3648	1,218.500	8.00	813,889 44,72	51.0039	9.4075	161.0954	29.7137
Hooge	9301122	2006	17.000	Container Vessel	15.633	Container ship	HFO; MGO; MDO	22,736.90	2,084.90	6,568.5447	1,959.600	2,126.40	145,866,600	91.6968	14.2932	288.8936	45.0312
Langeness	9301134	2006	17.000	Container Vessel	15.633	Container ship	HFO; MGO; MDO	18,623.00	1,885.50	5,953.0128	1,750.510	687.28	101,238 077,6	101.2619	18.6274	319.6592	58.8021
Norderoog	9256315	2004	17.000	Container Vessel	15.633	Container ship	HFO; MGO; MDO	20,233.90	1,847.60	5,835.0096	2,014.200	2,524.90	130,993,508	91.3121	14.1045	288.3779	44.5443
Sjard	9303314	2007	17.500	Multi Purpose Vessel/Tweendecker	12.936	General cargo ship	HFO; MGO; MDO	6,571.00	427.10	1,567.4182	774.770	212.05	5,899,947.8	75.6506	8.4255	238.5357	26.5667
Süderoog	9256327	2005	17.000	Container Vessel	15.633	Container ship	HFO; MGO; MDO	78,901.00	6,511.60	20,529.9426	6,455.400	711.10	372,447,709	82.5287	17.4833	260.1988	55.1217
Wybelsum	9386976	2008	17.094	Container Vessel	15.633	Container ship	HFO; MGO; MDO	45,202.00	3,840.30	12,162.9412	4,182.500	2,264.30	349,314,682	84.9581	10.9938	269.0779	34.8194

Source: Data from Akram Akoel, Capt. Ing., Nautical Superintendent Briese Schiffahrts GmbH.

References

Abel-Koch, J., and K. Ullrich. 2020. Kurzfristiger Schock mit langfristiger Wirkung: Corona-Krise und internationale Wertschöpfungsketten. *KfW Economic Research. Fokus Volkswirtschaft* Vol. 309: 1–7.

ADR. 2019. *ADR European Agreement*, vol. I and II, Błonie: Buch-Car.

Ahn, S.-B. 2005. Container Tracking and Tracing System to Enhance Global Visibility. *Proceedings of the Eastern Asia Society for Transportation Studies* Vol. 5.

Akoel, A., and R.-K. Miler. 2019. Economic and Operational Impact of the MRV Implementation on Maritime Transport Processes. *WSB Journal of Business and Finance* Vol. 53, No. 1: 133–143.

ALG. 2001. *Future Trends on Container Handling Systems.* https://algnewsletter.com/maritime/future-trends-on-container-handling-systems (accessed: 22nd March 2021).

AMT Marine. 2021. https://amtmarine.ca (accessed: 12th April 2021).

AMT Marine Software. 2021. https://amtmarine.ca/Pages/18/SimpleStow_Overview (accessed: 12th April 2022).

APM Terminals. 2017. *Directors Report Strategy and Performance.* https://180209-apm-terminals-2017-annual-results-pages-from-ap-moller-maersk-2017-annua.pdf (accessed: 10th February 2021).

APM Terminals. 2021. *11 Major Container Terminal Operators in the World.* https://www.apmterminals.com (accessed: 26th April 2021).

APPIS. 2022. *Automatyka przemysłowa i systemy sterowania.* www.apiss.pl/produkty/kalkulator-zuycia-paliwa-dla-statkw/ (accessed: 2nd March 2022).

Aquarius Gem Shipping. 2021. *Ubezpieczenie ładunku w transporcie morskim, lądowym i lotniczym.* https://aquarius-gem.com/ubezpieczenia-23-pl.html (accessed: 20th March 2021).

AVANTE International Technology Inc. 2013. www.avantetech.com/products/shipping (accessed: 12th November 2013).

AVANTE International Technology Inc. 2021. www.avantetech.com/products/shipping/ (accessed: 12th April 2021).

AXSMarine. 2021. *Alphaliner Top 100.* https://alphaliner.axmarine.com (accessed: 13th March 2021).

Baltic Container Terminal. 2021. *BCT Gdynia Poland.* www.bct.gdynia.pl (accessed: 26th April 2021).

Baltic Maritime Outlook. 2006. *Goods Flows and Maritime Infrastructure in the Baltic Sea Region.* Hamburg: BMT Transport Solutions GmbH, Centre for Maritime Studies.

Balticon. 2010. *Funkcjonowanie oraz perspektywy rozwoju rynku przewozów kontenerowych w Polsce do roku 2015 dla Balticon SA.* www.maritime.com.pl/newsletter/dok/Raport_rynek_kontenerowy.pdf (accessed: 10th February 2021).

Bartosiewicz, A. 2013. *Rozwój konteneryzacji na świecie od końca XIX w. do czasów współczesnych.* Studia z Historii Społeczno-Polityczne. vol. XI, Łódź.

Bartosiewicz, A. 2020. *Transport morski kontenerów.* Łódź: Wydawnictwo Uniwersytetu Łódzkiego.

BBA Transport System. 2021. *INCOTERMS 2020—jak wyglądają znane reguły handlowe w nowym wydaniu?* https://bbats.pl/poradnik-specjalistyczny/incoterms-2020/ (accessed: 12th July 2021).

BCT. 2021. *The tariff of fees charged for services provided at the Baltic Container Terminal.* https://cdnweb.bct.gdynia.pl/taryfa-uslug-bct_2021.pdf (accessed: 10th October 2021).

Blajer, A. 2000. *Międzynarodowe Reguły Handlowe*. Gdańsk: ODDK.

BohailLeasing Co. 2021. www.bohaiholding.com/en (accessed: 12th July 2021).

British International Freight Association. 2021. *Containers Lost at Sea—2020 Update*. www.bifa.org/news/articles/2020/jul/containers-lost-at-sea-2020-update (accessed: 10th February 2021).

Brodecki, Z. 2009. *Prawo ubezpieczeń morskich*. Sopot: Lex.

Brzozowska, A., and R. Miler. 2017. Implementation of the Green Shipping Practices as an Element of the Maritime Transport Restructuring Processes. In *Contemporary Issues and Challenges of the Organization Management Process. Models—Implementation—Interrelation*, eds. A. Jaki and T. Rojek. Cracow: Foundation of the Cracow University of Economics.

Brzozowski, M. 2012. Morskie przewozy kontenerowe jako generator zmian. *Zeszyty Naukowe Politechniki Warszawskiej zeszyt* No. 84.

Bujak, A., M. Smolarek, and A. Orzeł. 2013. Telematics Innovations in Transport Safety. In *Activities of Transport Systems Telematics*, ed. J. Mikulski. Berlin—Heidelberg: Springer.

Bureau Voorlichting Binnenvaart. 2018. www.bureauvoorlichtingbinnenvaart.nl/inland-navigation-promotion/basic-knowledge/waterways (accessed: 11th February 2018).

Carbon Trust. 2010. *Carbon Footprinting—the Next Step to Reducing Your Emissions*. www.carbontrust.co.uk (accessed: 22nd March 2022).

CargoX. 2017. *Innovation Platform CargoX Smart Bill of Lading*. www.gospodarkamorska.pl/Stocznie,Offshore/statki-autonomiczne-rozwiazaniem-przyszlosci.html (accessed: 12th December 2017).

Chip.pl. 2017. *Autonomiczne statki towarowe*. www.chip.pl/2017/05/autonomiczny-statek-towarowy-zasilany-energia-elektryczna/ (accessed: 11th March 2021).

The ClarkSea Index. 2021. *The Heart Rate Monitor of the Shipping Industry*. https://shippingresearch.wordpress.com (accessed: 20th February 2021).

Clarksons. 2021. *Containers*. www.clarksons.com/services/broking/containers (accessed: 20th February 2021).

CMA-CGM. 2021. *CMA CGM. Key Figures: A Major Player in the World Economy*. www.cmacgm-group.com/en/group/at-a-glance/key-figures (accessed: 10th February 2021).

CMD-Construction. 2021. *Customization Makes Mutual Customization*. www.construction-cmd.com/concepts-of-foldable-containers (accessed: 10th February 2021).

Commission of the European Communities. 1992a. *Directorate—General for Transport-Dg VII. Green Paper—Towards Fair and Efficient Pricing in Transport. Policy Options for Internalising the Externalcosts of Transport in the European Union, COM(95)691*. https://europa.eu/documents/comm/green_papers/pdf/com95_691_en.pdf (accessed: 2nd March 2022).

Commission of the European Communities. 1992b. *Green Paper—The Impact of Transport on the Environment. A Community Strategy for Sustainable Mobility. COM(92) 46 Final*. https://op.europa.eu/pl/publication-detail/-/publication/98dc7e2c-6a66-483a-875e-87648c1d75c8/language-en (accessed: 2nd March 2022).

Commission of the European Communities. 1998. *White Paper—Fair Payment for Infrastructure Use: A Phased Approach to a Common Transport Infrastructure Charging Framework in the EU. COM(1998) 466 Final*. https://op.europa.eu/en/publication-detail/-/publication/ceccf466-59bd-46e6-a08b-972286cebdc6/language-en (accessed: 2nd March 2022).

Commission of the European Communities. 2001a. *A Sustainable Europe for a Better World: A European Union Strategy for Sustainable Development, COM(2001) 264 Final*. https://eur-lex.europa.eu/legal-content/EN/ALL/?uri=CELEX:52001DC0264 (accessed: 2nd March 2022).

Commission of the European Communities. 2001b. *White Paper—European Transport Policy for 2010: Time to Decide. COM(2001) 370 Final*. http://aei.pitt.edu/1187/1/transport_2010_wp_com_2001_370.pdf (accessed: 2nd March 2022).

Commission of the European Communities. 2009. *A Sustainable Future for Transport: Towards an Integrated, Technology-Led and User Friendly System, COM(2009) 279 Final*.

Commission of the European Communities. 2010. *Europe 2020 a Strategy for Smart, Sustainable and Inclusive Growth. COM (2010) Final*. https://eur-lex.europa.eu/legal-content/EN/TXT/PDF/?uri=CELEX:52010DC2020&from=PL (accessed: 2nd March 2022).

Commission of the European Communities. 2019. *Reflection Paper Towards a Sustainable EUROPE by 2030. COM(2019) 22*. https://ec.europa.eu/info/sites/default/files/rp_sustainable_europe_30-01_en_web.pdf (accessed: 2nd March 2022).

Computers & Industrial Engineering. 2016. Container specification. *Hapag-Lloyd, Hamburg* Vol. 83: 316–326, Online Publication: www.sciencedirect.com/science/article/abs/pii/S0360835215000649 (accessed: 22nd March 2021).

Container xChange. 2021. *Biggest Container Manufacturers of the World*. www.container-xchange.com/blog/container-manufacturers-new-built-and-used-containers/ (accessed: 10th February 2021).

Convention on Customs Treatment of Pool Containers used in International Transport 1994. 2021. www.prawo.pl/akty/dz-u-2001-53-559,16898802.html (accessed: 23rd April 2021).

Corbett, J.-J., J.-J. Winebrake, E.H. Green, P. Kasibhatla, V. Eyring, and A. Lauer. 2007. Mortality from Ship Emissions: A Global Assessment. *Environmental Science & Technology* Vol. 41, No. 24: 12–17.

Cosco Shipping Holdings Co. Ltd. 2020. *Annual Report 2019*. Hong Kong.

Cosco Shipping Lines Co. 2021. http://lines.coscoshipping.com/home/About/about/Profile (accessed: 7th March 2021).

The Council and the European Parliament. 2006. *Keep Europe Moving—Sustainable Mobility for Our Continent. Mid-Term Review of the European Commission's 2001 Transport—White Paper, COM(2006) 314 Final*. https://eur-lex.europa.eu/legal-content/EN/TXT/PDF/?uri=CELEX:52006DC0314&from=EN (accessed: 2nd March 2022).

Country and Port Level Liner Shipping Connectivity Index. 2021. https://porteconomicsmanagement.org/pemp/contents/part1/ports-and-container-shipping/country-port-level-liner-shipping-connectivity-index (accessed: 10th March 2021).

Dąbrowski, J. 2013. Koncepcja zrównoważonego rozwoju w polityce transportowej Unii Europejskiej. In *Współczesne problemy rozwoju lądowo-morskich systemów transportowych*, ed. J. Dąbrowski and T. Nowosielski. Gdańsk: Instytut Transportu i Handlu Morskiego.

Dąbrowski, J., A. Kaliszewski, and H. Klimek. 2013. *Orientacja logistyczna operatorów portowych na przykładzie BCT Bałtyckiego Terminalu Kontenerowego Spółka z o.o.* www.researchgate.net/publication/310801955_Orientacja_logistyczna_operatorow_portowych_na_przykladzie_BCT_Baltyckiego_Terminalu_Kontenerowego_Sp_z_oo (accessed: 22nd March 2021).

DCT. 2021. *Podręcznik użytkownika systemu awizacji samochodów ciężarowych e-Brama v. 1.1*, internal materials provided by the DCT Gdańsk, Poland.

DCT. 2022. *Instrukcja obsługi systemu operacyjnego NAVIS*. Deepwater Container Terminal Gdynia, Poland, internal materials provided by the DCT.

DP World. 2021a. *DP World Handles 71 Million TEU and Reports 1.0% Volume Growth in 2019*. www.dpworld.com/news/releases/dp-world-handles-71-million-teu-and-reports-10-volume-growth-in-2019/ (accessed: 26th April 2021).

DP World. 2021b. *Ports & Terminals.* www.dpworld.com/services/ports-and-terminals (accessed: 23rd March 2021).

Drewry. 2021. *World Container Index.* www.drewry.co.uk/world-container-index-assessed-by-drewry (accessed: 10th February 2021).

DVZ Deutsche Verkehrs Zeitung. 2021. *Ocean Network Express schreibt erneut schwarze Zahlen.* www.dvz.de/rubriken/see/detail/news (accessed: 10th February 2021)

EC Regulation. 2004. *Regulation (EC) no. 725/2004 of the European Parliament and of the Council of 31st March 2004 on Enhancing Ship and Port Facility Protection.* https://eur-lex.europa.eu/legal-content/EN/TXT/PDF/?uri=CELEX:32004R0725&rid=7 (accessed: 10th February 2022).

E-logistyka. 2021. https://e-logistyka.pl/combiterms/ (accessed: 29th March 2021).

The Environmental Ship Index (ESI). 2022. www.environmentalshipindex.org (accessed: 2nd March 2022).

EU Ports European Economic Interest Group. 2017. *Guidance for Greenhouse Gas Emission Footprinting for Container Terminals.* December 2017. https://navclimate.pianc.org/news/download/21_9cd57e0cb26d8f27913e702868f8828e (accessed: 8th March 2022).

EUROACTIV. 2022. www.euractiv.com/climate-environment/brussels-launch-shipping-emissio-news-515108 (accessed: 2nd March 2022).

Eurokai. 2017. *The Eurokai Annual Report 2017.* www.eurokai.de/eurokai_en/Investor-Relations/Financial-Reports (accessed: 10th February 2021).

Eurokai. 2019. *The Eurokai Annual Report 2019.* www.eurokai.de/eurokai_en/Investor-Relations/Financial-Reports (accessed: 10th February 2021).

Eurokai Hauptversammlung. 2021. *Hauptversammlung der Eurokai.* www.eurokai.de/content/download/9721/107707/version/2/file/Einladung+Hauptversammlung+mit+Tagesordnung+2021.pdf (accessed: 16th February 2021).

Europa Systems. 2021. *ATLS—automatyczny załadunek i rozładunek tirów.* https://europa-systems.pl/product-pol-190-Automatyczny-zaladunek-kontenera-SkateLoader-System.html (accessed: 22nd March 2021).

European Commission. 2011. *Road map to a Single European Transport Area—Towards a Competitive and Resource Efficient—White Paper. COM(2011) 144 Final.* https://ec.europa.eu/transparency/documents register/api/files/COM(2011)144_0/de00000000 615091?rendition=false (accessed: 2nd March 2022).

European Economic and Social Committee. 2020. *Sustainable Development Observatory Work Programme 2020–2030.* www.eesc.europa.eu/sites/default/files/files/eesc-2020-05649-00-01-info-tra-en.pdf (accessed: 1st March 2022).

European Maritime Safety Agency. 2022. www.emsa.europa.eu (accessed: 20th February 2022).

European Parliament and the Council. 2015a. *On the Monitoring, Reporting and Verification of Carbon Dioxide Emissions from Maritime Transport, and Amending Directive 2009/16/EC.* https://eur-lex.europa.eu/legal-content/EN/TXT/PDF/?uri=CELEX:3201 5R0757&from=EL (accessed: 2nd March 2022).

European Parliament and the Council. 2015b. *Rozporządzenie Parlamentu Europejskiego i Rady UE w sprawie monitorowania, raportowania i weryfikacji emisji dwutlenku węgla z transportu morskiego oraz zmiany dyrektywy.* 2009/16/WE, Dz. Urz. UE L 123, 19.5.2015.

European Parliament and the Council. 2016. *Relating to a Reduction in the Sulphur Content of Certain Liquid Fuels (Codification). 2016/802.* https://eur-lex.europa.eu/legal-content/EN/TXT/PDF/?uri=CELEX:32016L0802&from=pl (accessed: 2nd March 2022).

Euro Shipping. 2021. www.euro-shipping.com.pl/?page_id=425 (accessed: 25th March 2021).

Eurostat ITF. 2021. Economic Commission for Europe Eurostat ITF (International Transport Forum), *Ilustrowany słownik statystyk transportu*. https://ec.europa.eu/eurostat/ramon/coded_files/transport_glossary_4_ed_PL.pdf (accessed: 26th April 2021).

Evergreen. 2021. *Evergreen Marine Corp. Corporate Profile*. www.evergreen-marine.com/tbi1/jsp/TBI1_CorporateProfile.jsp (accessed: 10th February 2021).

Federal Reserve Bank of Dallas. 2010. *Shipping Indexes Signal Global Economic Trends*, Globalization and Monetary Policy Institute, 2010 Annual Report. www.dallasfed.org/assets/documents/institute/annual/2010/annual10e.pdf (accessed: 10th October 2021).

Fertsch, M., ed. 2006. *Słownik terminologii logistycznej*. Poznań: Instytut Logistyki i Magazynowania.

Ficoń, K. 2013. *Logistyka morska. Statki, porty, spedycja*. Warszawa: Bel Studio.

Fitzmaurice, M. 2016. The International Convention for the Prevention of Pollution from Ships (MARPOL). In *IMLI Manual on International Maritime Law, Vol. III Marine Environmental Law and Maritime Security Law*, eds. D. Joseph Attard, Vol M. Fitzmaurice, N. A. Martínez Gutiér, and R. Hamza, 55–63. Oxford: Oxford University Press.

Florens. 2021. *Florens Asset Management Company Limited*. www.florens.com (accessed: 12th July 2021).

Forschungs Informations System. 2021. *Strategische und operative Gründefüf Leer container ungleichgewichte*. www.forschungsinformationssystem.de/servlet/is/381085/ (accessed: 10th February 2021).

Freightos Baltic Index. 2021. *FBX: Global Container Freight Index*. https://fbx.freightos.com (accessed: 10th February 2021).

Freight Transport Association. 2011. Heriot-Watt University, Edinburgh (McKinnon, A. and R. Woolford, Logistics Research Centre) and EPSRC. *Decarbonising the Maritime Supply Chain: The Shippers Perspective*. Workshop held in London 15 February 2011. https://decarbonisingfreight.co.uk/wp-content/uploads/2021/04/McKinnon-Woolford-2011-The-effects-of-port-centric-logistics-on-the-carbon-intensity-of-the-maritime-supply-chain.pdf (accessed: 25th March 2021).

Garrat, M., and Ch. Rowlands. 2011. Beyond the Port Fairways: Trends in the Carbon Footprint of the Deep-Sea Container Shipping Industry. *Green Port Magazine*, Summer 2011. www.greenport.com (accessed: 22nd March 2022).

Global Trade Magazine. 2021. www.globaltrademag.com/doubling-cars-shipping-containers/ (accessed: 10th February 2021).

Gort, R. 2009. *Design of an Autonomous Loading & Unloading Inland Barge. A Concept for Container Transport on the Albert Canal*. Antwerp: Delft.

Green Efforts. 2021. https://greenefforts.eu/wp-content/uploads/2019/02/D.3.2-GreenEFFORTS-Container-Terminal-Process.pdf (accessed: 13th March 2021).

Grulkowski, S., and J. Zariczny. 2012. Analiza celowości i możliwości budowy „suchego portu" w pobliżu Trójmiasta. *Technika Transportu Szynowego* No. 9.

Grzelakowski, A.-S. 2013. Rozwój globalnego handlu i systemu logistycznego i ich wpływ na rynek morskich przewozów kontenerowych. *International Business and Global Economy* Vol. 32.

Grzelakowski, A.-S. 2019. *Global Container Shipping Market development and its Impact on Mega Logistics System*. www.researchgate.net/publication/335490990_Global_Container_Shipping_Market_Development_and_Its_Impact_on_Mega_Logistics_System (accessed: 20th February 2021).

Grzelakowski, A.-S., and M. Matczak. 2013. *Współczesne porty morskie. Funkcjonowanie i rozwój*. Gdynia: Akademia Morska w Gdyni.

Grzywacz W., and J. Burnewicz. 1989. *Ekonomika transportu*. Warszawa: Wydawnictwa Komunikacji i Łączności.

Grzybowski, L., B. Łączyński, A. Narodzonek, and J. Puchalski. 1997. *Kontenery w transporcie morskim*. Gdynia: Trademar.

Hapag-Lloyd. 2020a. *Annual Report 2019*, Hamburg.

Hapag-Lloyd. 2020b. *Geschäftsbericht 2019*, Hamburg.

Hapag-Lloyd. 2021. *About Us*. www.hapag-lloyd.com/en/about-us.html#anchor_171183 (accessed: 10th February 2021).

Harpex. 2021a. https://harpex.harperpetersen.com/harpex-concept.html (accessed: 20th February 2021).

Harpex. 2021b. https://ycharts.com/indicators/reports/harpex (accessed: 10th February 2021).

Harrison, J. 2016. Atmospheric Pollution of the Marine Environment. In *International Maritime Law, Vol. III Marine Environmental Law and Maritime Security Law*, ed. D.-J. Attard. Oxford: Oxford University Press.

Hartman, B. 2009. Subsidiarity in EU Environmental Policy. *Problems of Sustainable Development* Vol. 4, No. 1: 93–98.

Hebel, A. 2005. *Poradnik ubezpieczeń morskich*. Wrocław: Oficyna Wydawnicza Foka.

Hellenic Shipping News. 2021. *'Exceptional Market' Conditions in Shipping to Persist Until Q1 2022, Says Maersk*. www.hellenicshippingnews.com/exceptional-market-conditions-in-shipping-to-persist-until-q1-2022-says-maersk/ (accessed: 4th November 2021).

Hermanowski, J. 2004. *Incoterms*. Warszawa—Zielona Góra: UNIVERS.

Hoffmann, J., and J. Hoffmann. 2020. Ports in the Global Liner Shipping Network: Understanding Their Position, Connectivity, and Changes Over Time. *Transport and Trade Facilitation Newsletter no. 87—Third Quarter 2020*, UNCTAD No. 5.

Hutchison Ports Busan. 2021. www.hktl.com/sub02/02.php?catetop=0202&lang=eng (accessed: 22nd March 2021).

ICHCA International. 2020. *The International Maritime Dangerous Goods (IMDG) Code. Briefing Pamphlet, #3 9th Edition*. www.bifa.org/media/4544564/bp-3-the-international-maritime-dangerous-goods-imdg-code.pdf (accessed: 2nd March 2022).

ICTSI Annual Report. 2020. *International Container Terminal Systems Inc.* Annual Reports. www.ictsi.com/reports-and-presentations/annual-reports (accessed: 10th February 2022).

IFEU Heidelberg. 2011. Okö-Institut, IVE and RMCON. *EcoTransIT—World Ecological Transport Information Tool for Worldwide Transports, Methodology and Data Update*. Commissioned by DB Schenker Germany and UIC (International Union of Railways). Berlin/Hanover/Heidelberg. 31 July 2011.

Igliński, H. 2012. Wybrane sposoby ograniczenia zużycia paliwa w transporcie morskim. *Logistyka* No. 5.

IMDG Code. 2018. *International Maritime Dangerous Goods Code*. London: IMO Publishing.

IMO. 2021. www.imo.org/OurWork/Environment/PollutionPrevention/AirPollution/Pages/GHG-Emissions.aspx (accessed: 25th March 2021).

IMO CSC. 1972. *Convention for Safe Containers (CSC) 1972, Interpretations and Guideleines, IMO*. www.imo.org/en/About/Conventions/Pages/International-Convention-for-Safe-Containers-(CSC).aspx (accessed: 26th February 2021).

IMO/ILO/UNECE Code of Practice. 2014. *Code of Practice for Packing of Cargo Transport Units (CTU Code)*. International Maritime Organization, MSC.1/Circ.1497, 16 December 2014, London.

IMO/ILO/UNECE Guidelines. 1997. *Guidelines for Packing of Cargo Transport Units (CTUs)*. International Maritime Organization, MSC/Circ.787 2nd May 1997, London.

INFOCar.net. 2021. *Transport Multimodalny*. https://infocar.net.pl/transport-multimodalny-co-to-jest-i-jakie-sa-jego-zalety.html (accessed: 26th April 2021).

Ingenieur.de. 2021. *Die größten Containerschiffe der Welt.* www.ingenieur.de/technik/ fachbereiche/schiffbau/die-groessten-containerschiffe-der-welt (accessed: 12th February 2021).

Institute of Chartered Shipbrokers. 2021. www.ics.org.uk/media/628894/new%20contex%20 commentary%202021%20week%2001.pdf (accessed: 20th March 2021).

Intergovernmental Organisation for International Carriage by Rail. n.d. *Regulation concerning the International Carriage of Dangerous Goods by Rail (RID).* www.otif. org/fileadmin/user_upload/otif_verlinkte_files/07_veroeff/03_erlaeut/rpex99-rid-e.pdf (accessed: 10th February 2021).

International Association of Ports and Harbors. 2022. *World Port Sustainability Program.* https://sustainableworldports.org (accessed: 2nd March 2022).

International Federation of Freight Forwarders Associations. 2010. *Documents and Forms.* https://fiata.org/fileadmin/user_upload/documents/Diverses/FIATA_Documents_and_ Forms.pdf (accessed: 2nd March 2022).

International Maritime Organization. 1972. *The International Convention for Safe Containers 1972.* www.imo.org/en/About/Conventions/Pages/International-Convention-for-Safe-Containers-(CSC).aspx (accessed: 10th Feb. 2021).

International Maritime Organization. 1978. *Annex II—Regulations for the Control of Pollution by Noxious Liquid Substances in Bulk, Regulation 6 Categorization and Listing of Noxious Liquid Substances and Other Substances.* www.marpoltraining. com/MMSKOREAN/MARPOL/Annex_II/r6.htm (accessed: 2nd March 2022).

International Maritime Organization. 1992. *Annex III—Regulations for the Prevention of Pollution by Harmful Substances Carried by Sea in Packaged Form, Regulation 4.* www.marpoltraining.com/MMSKOREAN/MARPOL/Annex_III/r4.htm (accessed: 2nd March 2022).

International Maritime Organization. 1993. *International Management Code for the Safe Operation of Ships and for Pollution Prevention (International Safety Management ISM Code).* IMO Res. A.741(18), amended: MSC.104(73), MSC.179(79), MSC.195(80), MSC.273(85) and MSC.359(92). https://wwwcdn.imo.org/localresources/en/KnowledgeCentre/ IndexofIMOResolutions/AssemblyDocuments/A.741(18).pdf (accessed: 2nd March 2022).

International Maritime Organization. 1996. *Annex I—Regulations for the Prevention of Pollution by Oil. Chapter 4—Requirements for the Cargo Area of Oil Tankers, Part A— Construction, Regulation 19.* www.marpoltraining.com/MMSKOREAN/MARPOL/ Annex_I/r19.htm (accessed: 2nd March 2022).

International Maritime Organization. 2000. Amendments to the Revised Specifications for the Design, Operation and Control of Crude Oil Washing Systems (Resolution A.446 (XI), as amended by Resolution A.497(XII) Resolution A.897(21), A 21/Res.8974. https:// wwwcdn.imo.org/localresources/en/KnowledgeCentre/IndexofIMOResolutions/ AssemblyDocuments/A.897(21).pdf (accessed: 2nd March 2022).

International Maritime Organization. 2004. *International Convention for the Control and Management of Ships' Ballast Water and Sediments.* www.aph.gov.au/parliamentary_ business/committees/house_of_representatives_committees?url=jsct/13june2007/ treaties/ships_text.pdf (accessed: 2nd March 2022).

International Maritime Organization. 2011. *Technical and Operational Measures to Improve the Energy Efficiency of International Shipping and Assessment of Their Effect on Future Emissions* (SBSTA 35). https://seors.unfccc.int/applications/seors/attachments/get_attachm ent?code=MOE2AHKBQJAQLFCWO8EAVIYFKOE25OT4 (accessed: 2nd March 2022).

International Maritime Organization. 2012. *International Convention for the Prevention of Pollution from Ships, 1973.* Appendix III, p. 129–130. http://library.arcticportal. org/1699/ (accessed: 2nd March 2022).

International Maritime Organization. 2017. *International Convention for the Prevention of Pollution from Ships* (MARPOL). https://wwwcdn.imo.org/localresources/en/publications/Documents/Supplements/English/QQQE520E_092020.pdf (accessed: 2nd March 2022).

International Maritime Organization. 2021. www.imo.org/en/OurWork/Safety/Pages/Dangerous Goods-default.aspx (accessed: 16th February 2021).

IPSI Final Report. 2008. *Improved Port/ship Interface (IPSI)*, Deliverable D 6005 IPSI WP6200, Final Report for Publication.

ISL Shipping Statistics and Market Review. 2020. (SSMR) (charter rates), ISL Institute of Shipping Economics and Logistics, May/June 2020, Bremen, Germany.

ISO. 2013. *Series 1 Freight Containers—Classification, Dimensions and Ratings. International Standard ISO 668*, ISO. https://www.iso.org/standard/59673.html (accessed: 9th February 2022).

ISO 1496–3. 2019. *Series 1 Freight Containers—Specification and Testing—Part 3: Tank Containers for Liquids, Gases and Pressurized Dry Bulk*, International standard ISO 1496–3 Fifth edition 2019–04. www.iso.org/obp/ui/#iso:std:iso:1496:-3:ed-4:v1:en:sec:B (accessed: 9th February 2021).

ISO 14064–1. 2006. *International Standards Organisation, Specification with Guidance at the Organisational Level for Quantification and Reporting of GHG Emissions and Removals*. www.iso.org (accessed: 22nd March 2022).

Jakowski, S. 2017. *Ochrona ładunków w przewozach kontenerowych*. Warszawa: Wydawnictwa Komunikacji i Łączności.

Janicka, B. 2012. FIATA Multimodal Transport Bill of Lading czyli spedytor przewoźnikiem. *Biuletyn Polskiej Izby Spedycji i Logistyki* No. 11/12.

Jarysz-Kamińska, E. 2013. Ocena ryzyka w transporcie morskim. *Logistyka* No. 6.

Journal of Laws. 1972. *International Convention on Safe Containers*, Geneva, 2nd December 1972, 1984.24.118, Art. IV. http://www.admiraltylawguide.com/conven/containers1972.html (accessed: 9th May 2022).

Journal of Laws. 2001. Act of 6th September 2001 on Road Transport. *Journal of Laws* No. 125, item 1371.

Journal of Laws. 2008. Act of 4th September 2008 on Protection of Shipping and Sea Ports, *Journal of Laws* No. 692.

Journal of Laws. 2018a. The Act of 18th September 2001. *The Maritime Code*, 2018.0.2175, Art. 136.

Journal of Laws. 2018b. The Act of 18th September 2001. *The Maritime Code*, 2018.0.2175, Aart. 139. 1.

Journal of Laws. 2018c. Act of 18th September 2001 on Maritime Code. *Journal of Laws* 2018.0.2175, Art. 101. 2.

Jurdziński, M. 2012a. Innowacje technologiczne na statkach morskich w celu redukcji zużycia energii i emisji CO_2. *Zeszyty Naukowe Akademii Morskiej* No. 77.

Jurdziński, M. 2012b. *Technological Innovations on Sea-Going Ships to Reduce Energy Consumption and CO_2 Emissions*. Gdynia: Maritime Academy.

Kasprzyk, M. 2019. *Znowelizowane Amerykańskie Definicje w Handlu Zagranicznym*. https://mfiles.pl/pl/index.php/Znowelizowane_Amerykańskie_Definicje_w_Handlu_Zagranicznym (accessed: 26th April 2019).

Kaup, M., and M. Chmielewska-Przybysz. 2012. Analiza i ocena technologii wykorzystywanych do obsługi kontenerów w portach morskich. *Logistyka* No. 5.

Khan, F-I., and A-K. Ghoshal. 2000. Removal of Volatile Organic Compounds from Polluted Air. *Journal of Loss Prevention in the Process Industries* Vol. 13: 527–545.

Kiliński, R. 2018. Tajniki ubezpieczenia armatora morskiego. *Miesięcznik Ubezpieczeniowy* No. 6.

Klose, A. 2012. *20 Fuß Äquivalent Einheit. Die Herrschaft der Containerisierung.* www. containerwelt.info/pdf/Abstract_Container_wisssch.pdf (accessed: 2nd February 2012).

Kostrzewski, A., M. Nader, and M. Kostrzewski. 2018. Racjonalizacja rozłożenia wybranych jednostek transportu intermodalnego na długości ładunkowej pociągu. *Prace Naukowe Politechniki Warszawskiej* Vol. 120.

Krasucki, Z., and J. Neider. 1986. *Konteneryzacja w transporcie międzynarodowym.* Warszawa: Państwowe Wydawnictwo Naukowe.

Książkiewicz, D., and D. Mierkiewicz. 2014. Techniczne i organizacyjne aspekty bezpieczeństwa w morskich przewozach kontenerowych. *Logistyka* No. 4.

Kubicki, J., I. Urbanyi-Popiołek, and J. Miklińska. 2002. *Transport międzynarodowy i multimodalne systemy transportowe.* Gdynia: Fundacja Rozwoju Wyższej Szkoły Morskiej.

Kubowicz, D. 2019. Zarządzanie procesami przepływu ładunków na morskim terminalu kontenerowym z wykorzystaniem systemów informatycznych typu TOS. *Autobusy* No. 1–2.

Kuhne+Nagel. 2021. *Poprawne mocowanie kontenerów.* https://pl.kuehne-nagel.com/docum ents/244237/338584/2020+poprawne+mocowanie+kontener%C3%B3w.pdf/42209ce2-9c1a-2483-d68d-ac51f54b088c?t=1599219755949 (accessed: 20th February 2021).

Kujawa, J., ed. 2015. *Organizacja i technika transportu morskiego.* Gdańsk: Wydawnictwo Uniwersytetu Gdańskiego.

Kujawa, J. 2020. *Ekonomika transportu morskiego i polityka żeglugowa.* Gdańsk: Wydawnictwo Uniwersytetu Gdańskiego, 2020.

Kulczycka, J., and M. Wernicka. 2015. Zarządzanie śladem węglowym w przedsiębiorstwach sektora energetycznego w Polsce—bariery i korzyści. *Polityka Energetyczna—Energy Policy Journal* Vol. 18, No. 2: 61–72.

Kulczyk, J., and J. Winter. 2003. *Śródlądowy transport wodny,* Wrocław 2003. www.dbc. wroc.pl/Content/1322/srodladowy_transport_wodny.pdf (accessed: 26th April 2021).

Kunert, J. 1970. *Technika handlu morskiego.* Warszawa: Państwowe Wydawnictwo Ekonomiczne.

Kurek, A., and T. Ambroziak. 2017. Wybrane zagadnienia pozycjonowania pustych kontenerów przy udziale transportu kolejowego. *Prace Naukowe Politechniki Warszawskiej. Transport* No. 117.

Lindstad, H., and E. Uthaug. 2003. *Innovative Technology: Ro-Ro Vessel, Terminal, and Barge Design That Will Improve the Cost Position and Lead-Time for the Logistic Chain.* IMDC 03 Conference proceedings, Athens, 3–5 June, 2003.

Lloyd's. 2021. https://lloydslist.maritimeintelligence.informa.com/LL1135004/Top-10-box-port-operators-2020 (accessed: 16th February 2021).

Łopuski J., ed. 1998. *Prawo morskie.* Bydgoszcz: Oficyna Wydawnicza Branta.

Lundkvist, M. 2010. *Simplified Model of Accident Consequences That Can Result from Collision, Contact, and Grounding Accidents* (origin: Lundkvist, M. 2010. Risk värdering av sjötrafik information). Draft report 2010–12–30. Swedish Maritime Administration.

Maersk. 2020. *Maersk Annual Report.* https://ml-eu.globenewswire.com/Resource/Download/ 35da68a1-f018-4dc3-85f9-95994bc5dc6b (accessed: 22nd March 2021).

Maersk. 2021a. www.maersk.com/local-information/europe/poland/import (accessed: 22nd March 2022).

Maersk. 2021b. www.maersk.com/local-information/europe/poland/export (accessed: 22nd March 2022).

Malinowska, M. 2016. *Armatorzy schodzą na ląd. Czy będą zagrożeniem dla spedycji?* www. gospodarkamorska.pl/Porty,Transport/armatorzy-schodza-na-lad-czy-beda-zagrozeniem-dla-spedycji.html (accessed: 6th March 2020).

Marciniak-Neider, D. 2008. *Płatności w handlu zagranicznym.* Gdańsk: Wydawnictwo Uniwersytetu Gdańskiego.

Marciniak-Neider, D. and J. Neider, eds. 2014. *Podręcznik spedytora*. Gdynia: Polska Izba Spedycji i Logistyki.

Marek, R. 2017. Korzyści ekonomiczne eksploatacji kontenerów składanych dla uczestników kontenerowego łańcucha logistycznego. *Studia i Materiały Instytutu Transportu i Handlu Morskiego* No. 14.

The Marine Environment Protection Committee. 2011. *Amendments to the Annex of the Protocol of 1997 to Amend the International Convention for the Prevention of Pollution from Ships, 1973, as Modified by the Protocol of 1978 Relating Thereto*. MEPC.203(62). https://wwwcdn.imo.org/localresources/en/OurWork/Environment/Documents/Technical%20and%20Operational%20Measures/Resolution%20MEPC.203%2862%29.pdf (accessed: 2nd March 2022).

The Marine Environment Protection Committee. 2013. *Guidelines for Calculation of Reference Lines for Use with the Energy Efficiency Design Index* (EEDI). Res.MEPC.231(65). https://wwwcdn.imo.org/localresources/en/KnowledgeCentre/IndexofIMOResolutions/MEPCDocuments/MEPC.231(65).pdf (accessed: 2nd March 2022).

The Marine Environment Protection Committee. 2014a. *Guidelines on the Method of Calculation of the Attained Energy Efficiency Design Index (EEDI) for New Ships defined the default emission factors for various types of fuels used in Maritime Transportation*. Resolution MEPC.245(66).

The Marine Environment Protection Committee. 2014b. *Standard specification for shipboard incinerators*. Resolution MEPC.244(66). https://wwwcdn.imo.org/localresources/en/KnowledgeCentre/IndexofIMOResolutions/MEPCDocuments/MEPC.244(66).pdf (accessed: 2nd March 2022).

The Marine Environment Protection Committee. 2016a. *Amendments to the Annex of the Protocol of 1997 to Amend the International Convention for the Prevention of Pollution from Ships, 1973, as Modified by the Protocol of 1978 Relating Thereto*. MEPC.278(70). https://wwwcdn.imo.org/localresources/en/OurWork/Environment/Documents/278(70).pdf (accessed: 2nd March 2022).

The Marine Environment Protection Committee. 2016b. *Guidelines for the Development of a SHIP ENERGY EFFICIENCY MANAGEMENT PLAN (SEEMP)*. Res.MEPC.282(70). https://wwwcdn.imo.org/localresources/en/KnowledgeCentre/IndexofIMOResolutions/MEPCDocuments/MEPC.282(70).pdf (accessed: 2nd March 2022).

The Marine Environment Protection Committee. 2018a. *Guidelines on the Method of Calculation of the Attained Energy Efficiency Design Index (EDDI) for New Ships.* (Res.MEPC.308(73). https://wwwcdn.imo.org/localresources/en/KnowledgeCentre/IndexofIMOResolutions/MEPCDocuments/MEPC.308(73).pdf (accessed: 2nd March 2022).

The Marine Environment Protection Committee. 2018b. *Initial IMO Strategy on Reduction of Ghg Emissions from Ships*. Res.MEPC.304(72). https://wwwcdn.imo.org/localresources/en/KnowledgeCentre/IndexofIMOResolutions/MEPCDocuments/MEPC.304(72).pdf (accessed: 2nd March 2022).

The Marine Environment Protection Committee. 2019. *Guidelines on Survey and Certification of the ENERGY EFFICIENCY DESIGN INDEX (EEDI)*, as amended. Wytyczne przeglądów i certyfikacji projektowego wskaźnika efektywności energetycznej EEDI, MEPC.261(68) oraz MEPC.309(73), wersja skonsolidowana w formie okólnika MEPC.1/Circ.855/Rev.2(Res.MEPC.254(67). https://wwwcdn.imo.org/localresources/en/OurWork/Environment/Documents/MEPC.1-Circ.855-Rev.2.pdf (accessed: 2nd March 2022).

Marine Link. 2021. www.marinelink.com/news/average-drewry-rates420386 (accessed: 10th February 2021).

Markusik, S. 2013. *Infrastruktura logistyczna w transporcie*, vol. II. Gliwice: Wydawnictwo Politechniki Śląskiej.

MDST. 2022. *MDS Transmodal*. https://www.mdst.co.uk/ (accessed: 18th April 2021).

Meadows, D.-H., J. Randers, and W.-W. Behrens III. 1972. *The Limits to Growth*. New York: Universe Books.

Miler, R. 2011. Systemy zobrazowania żeglugi i aplikacje informatyczne transportu morskiego. In *Wyzwania i problemy transportu morskiego*, ed. J. Dąbrowski. Gdańsk: Foundation for the Development of the University of Gdańsk.

Miler, R. 2015. Electronic Container Tracking Systems a Cost-Effective Tool in Intermodal and Maritime Transport Management. *Economic Alternatives* No. 1.

Miler, R. 2016a. *Bezpieczeństwo transportu morskiego*. Warszawa: PWN.

Miler, R. 2016b. *Monitoring bezpieczeństwa transportu morskiego—modelowanie systemów—strategie ekonomizacji*. Gdańsk: Wydawnictwo Uniwersytetu Gdańskiego.

Miler, R. 2019. *Telematyka w zarządzaniu transportem wodnym*. Warszawa: PWN.

Miler, R., and A. Bujak. 2014. Implementacja systemów monitoringu zużycia paliw i efektywności energetycznej statku jako pochodna dyrektyw proekologicznych UE w transporcie morskim. *Logistyka* No. 3: 4391–4399.

Miler, R., and A. Kuriata. 2019. The Conceptual Algorithm of the International Multimodal Transport (CAIMT). *Logistics and Transport* Vol. 1, No. 41.

Miler, R., and T. Szczepaniak. 2014. EU's GHGs from International Shipping Policy Impact on BSR Seaborne Trade Competition. *Prace Naukowe Wyższej Szkoły Bankowej w Gdańsku*: 93–111.

Mindur, L., ed. 2004. *Współczesne technologie transportowe*. Radom: Wydawnictwo Politechniki Radomskiej.

Miotke-Dzięgiel, J. 1996. *Morskie przewozy kontenerowe*. Gdańsk: Wydawnictwo Uniwersytetu Gdańskiego.

Misztal, K., ed. 2010. *Organizacja i funkcjonowanie portów morskich*. Gdańsk: Wydawnictwo Uniwersytetu Gdańskiego.

Misztal, K., L. Kuźma, and S. Szwankowski. 1994. *Organizacja i eksploatacja portów morskich*. Gdańsk: Wydawnictwo Uniwersytetu Gdańskiego.

Młynarczyk, J. 1997. *Prawo morskie*. Gdańsk: Info Trade.

Moret, C., and A. Lane. 2016. Port Performance and Productivity. Lifting the Game Part 2. *Baltic Transport Journal* Vol. 1.

MSC. 2022. www.msc.com/country-guides/poland/export-local-requirements/storage (accessed: 22nd March 2022).

MSC Mediterranean Shipping Company. 2021. *About Us*. www.msc.com/bih/about-us (accessed: 6th March 2021).

mySped Worldwide Logistics. 2021. *Ubezpieczenie cargo w transporcie*. http://mysped.pl/pl/transport-kontenerow-ubezpieczenie-cargo-w-transporcie/ (accessed: 10th February 2021).

Neider, J. 2012. *Transport międzynarodowy*. Warszawa: PWE.

Neider, J., and D. Marciniak-Neider. 1995. *Przewozy intermodalne w handlu międzynarodowym*. Warszawa: Państwowe Wydawnictwo Naukowe.

Neider, J., and D. Marciniak-Neider. 1997. *Transport Intermodalny*. Warszawa: Państwowe Wydawnictwo Ekonomiczne.

New CoTex. 2021. www.vhbs.de/index.php?id=28&L=1 (based on International Convention for Safe Containers) (accessed: 2nd March 2021).

Nicholls Y. 1973. *Emergence of Proposals for Recompensing Developing Countries for Maintaining Environmental Quality*. Switzeland: International Union for Conservation of Nature and Natural Resources Morges.

Nierzwicki, W., M. Richert, M. Rutkowska, and M. Wiśniewska. 1997. *Opakowania. Wybrane zagadnienia*. Gdynia: Wydawnictwo Wyższej Szkoły Morskiej w Gdyni.

Notteboom, T., A. Pallis, and J.-P. Rodrigue. 2022. *Port Economics, Management and Policy*. New York: Routledge.

Nowak, I. 2019. Co będzie kształtować morski transport kontenerów w najbliższej przyszłości. *Logistyka* No. 1.

Nowosielski, T. 2017. Ewolucja funkcjonowania przedsiębiorstw żeglugi kontenerowej w łańcuchach logistycznych. In *Przedsiębiorstwo w łańcuchu dostaw*, eds. E. Ignaciuk and J. Dąbrowski. Gdańsk: Polskie Towarzystwo Ekonomiczne.

Official Journal of the European Communities. 1997. *Treaty of Amsterdam Amending the Treaty on European Union, the Treaties Establishing the European Communities and Certain Related Acts* (97/C 340/01). https://eur-lex.europa.eu/legal-content/EN/TXT/?uri=celex%3A11997D%2FTXT (accessed: 2nd March 2022).

Official Journal of the European Union. 2009. *Directive 2009/16/EC sig* (MRV)—L 123/57. www.imo.org/MediaCentre/HotTopics/GHG/Pages/default.aspx (accessed: 20th October 2021).

ONE Ocean Network Express. 2021. *Corporate Factsheet.* www.one-line.com/sites/g/files/lnzjqr776/files/2020-01/ONE-Corporate-Factsheet%20%28updated%2020200115%29.pdf (accessed: 10th February 2021)

Opakowania.com.pl. 2021. https://opakowania.com.pl/news/rodzaje-workow-typu-big-bag-64187.html (accessed: 16th February 2021).

Opus Terminal. 2021. www.g2.com/products/opus-terminal-m/reviews (accessed: 30th March 2021).

Pacific International Lines. 2021. www.pilship.com/en-pil-today/150.htm (accessed: 13th March 2021).

Pamel, P-G. 2011. *Bills of Lading vs Sea Waybills, and The Himalaya Clause*, presented at the NJI/CMLA, Federal Court and Federal Court of Appeal Canadian Maritime Law Association Seminar April 15, 2011. Ottawa: Fairmont Château Laurier.

Paradigma Unternehmensberatung GmbH. 2021. *Interner Preismechanismen für leere Container.* www.paradigma.net/index.php/de/77-loesungen/management-loesungen/591-empty-container-repositioning-de (accessed: 26th April 2021).

Pawlik, T. 1999. *Seeverkehrswirtschaft. Internationale Containerlinienschifffahrt Eine betriebswirtschaftliche Einführung.* Wiesbaden: Gabler Verlag Springer Fachmedien.

Piskozub, A. 1982. *Gospodarowanie w transporcie.* Warszawa: WKŁ.

PN-ISO 830. 1999. *Freight Containers—Vocabulary.* https://www.iso.org/standard/1238.html (accessed: 2nd March 2022).

Polish Customs Department. 2021a. *Guidelines for the Treatment of Transport and Insurance Costs in relation with Defining the Customs Value of Imported Goods.* Customs Policy Department of the Ministry of Finance. www.podatki.gov.pl/clo/informacje-dla-przedsiebiorcow/wartosc-celna/traktowanie-kosztow-transportu-i-ubezpieczenia-w-zwiazku-z-ustalaniem-wartosci-celnej-sprowadzanych-towarow (accessed: 20th October 2021).

Polish Customs Department. 2021b. *Information Materials on Problems Related to the Implementation and Export and the ECS (AES) System in the Light of Changes Resulting from the Implementation of the EU Customs Code.* www.podatki.gov.pl/media/6308/mataria%C5%82-infirmacyjny-dot-wywozu-w-kontek%C5%9Bcie-zmian-wynikaj%C4%85cych-z-wej%C5%9Bciem-w-%C5%BCycie-unijnego-kodeksu-celnego.pdf (accessed: 20th July 2021).

Polish Register of Shipping PRS. 2012a. *Przepisy budowy kontenerów.* Gdańsk.

Polish Register of Shipping PRS. 2012b. *Przepisy nadzoru kontenerów w eksploatacji.* Gdańsk.

Polish Register of Shipping PRS. 2012c. *Ograniczanie emisji toksycznych związków spalania paliw ze statków w rejonie M. Bałtyckiego i portów*, *Seminarium projektowe.* Gdańsk: Polski Rejestr Statków.

Polish Register of Shipping PRS. 2016. *Kontenery morskie, przepisy.* Publication no. 112/P, Gdańsk.

Polish Register of Shipping PRS. 2020. *Rules publication 58/p hull surveys of double hull oil tankers.* Gdańsk: Polski Rejestr Statków.

Polish Register of Shipping PRS. 2022. *Rules for statutory survey of sea-going ships, Part IX, Environmental protection.* Gdańsk: Polish Register of Shipping.

Poradnik Przedsiębiorcy. 2021. *Odprawa celna przy imporcie towarów—kwestie formalne.* https://poradnikprzedsiebiorcy.pl/-odprawa-celna-przy-imporcie-towarow-kwestie-formalne (accessed: 22nd March 2021).

Port of Gdynia Authority SA. 2021. *The Tariff of Port Fees.* www.port.gdynia.pl/files/port/taryfa/taryfa-oplat-portowych_2020.pdf (accessed: 20th March 2021).

Port of Gdynia and Maritime Services. 2009. *Tariff for Services of WUŻ—Gdynia Port and Maritime Services Ltd.* www.port.gdynia.pl/files/port/taryfa/wuz_taryfa_2009.pdf (accessed: 20th March 2021).

Port of Luka Koper. 2015. *Rethinking Container Management System, Terminal Description and Requirements for Luka Koper.* Ref. Ares (2015)3799142–14/09/2015.

Potts, T. 2018. Climate Change, Ocean Acidification and the Marine Environment. In *International Marine Environmental Law and Policy*, ed. D. Hassan and M. S. Karim. London: Taylor & Francis Group.

Prawo ochrony środowiska. 2001. Dz.U.2001.62.627, art. 3 ust. 50. https://isap.sejm.gov.pl/isap.nsf/download.xsp/WDU20010620627/U/D20010627Lj.pdf (accessed: 1st March 2022).

Przepisy klasyfikacji i budowy statków morskich. Część I Zasady klasyfikacji. 2021. Gdańsk: Polski Rejestr Statków. https://www.prs.pl/uploads/mor_p1.pdf (accessed: 28th April 2020).

PSA International. 2021. www.globalpsa.com/psa-international (accessed: 16th March 2021).

PSA International Sustainability Report. 2020. www.globalpsa.com/wp-content/uploads/PSA-International-Sustainability-Report-2020.pdf (accessed: 2nd March 2022).

Pyć, D. 2012. Uwagi de lege ferenda o statusie Morza Bałtyckiego jako obszaru kontroli emisji tlenków azotu ze statków morskich. *Prawo Morskie* Vol. XXVIII: 235–247.

Pyć, D. 2019. Techniczne i operacyjne środki efektywności energetycznej dla statków morskich. *Prawo Morskie* Vol. XXXVII: 106–117.

Ramphul, M., V. Ramesh, and V.-C. Jaunky. 2017. The Impact of Foldable/ Collapsible Containers on Empty Container Management: A Case Study at Port Louis. *International Journal of Management and Applied Science* Vol. 3, No. 11.

Rezolucja Parlamentu Europejskiego. 2010. *Zrównoważona przyszłość transportu* (2011/C 351 E/03). https://eur-lex.europa.eu/legal-content/PL/TXT/HTML/?uri=CELEX:52010IP0260&from=FI (accessed: 2nd March 2022).

Resolution MEPC.292(71) (2017). *Guidelines for Administration Verification of Ship Fuel Oil Consumption Data.* https://wwwcdn.imo.org/localresources/en/KnowledgeCentre/IndexofIMOResolutions/MEPCDocuments/MEPC.292(71).pdf (accessed: 19th April 2021).

Rodrigue, J.-P. 2017. *Dept. of Global Studies & Geography, Hofstra University*, New York. https://people.hofstra.edu/geotrans/eng/ch3en/conc3en/intermodal_transmodal.html (accessed: 26th April 2019).

RSD Container Industry Specialist. 2021. *Quadruple Containers (Quadcon).* www.rsdmachinery.com/products-en.asp?ProductID=177&ClassID=14 (accessed: 10th February 2021).

Rydzkowski, W., and K. Wojewódzka-Król. 2020. *Transport.* Warszawa: WN PWN.

Safety4Sea. 2022. https://safety4sea.com/fuel-management-system-upgraded-with-emissions-monitoring-tech/ (accessed: 2nd March 2022).

Sagarra, R. M., J. de Souza, J. Martin, and J. Rodrigo. 2009. *El transporte de contenedores: Terminales, operatividad y casuística*. Barcelona: Universitat Politècnica De Catalunya.

Salomon, A. 2017. *Sztauowanie ładunków jako istotny element modelowania multimodalnych łańcuchów transportowych (na przykładzie Portu Gdynia)*. Konferencja Naukowa Logistyka Morska. LogMare 2017, Jastarnia, www.akademor.webd.pl/download/salomon2017_sztauowanie_ladunkow.pdf (accessed: 8th February 2021).

SeaCube Container Leasing. 2021. https://seacubecontainers.com (accessed: 12th July 2021).

Second IMO GHG Study. 2009. London: International Maritime Organisation.

ShipHub. 2021b. *Karnet TIR*. www.shiphub.pl/karnet-tir/ (accessed: 22nd March 2021).

Shipping and Freight Resource. 2021. *Pallet wide Containers*. www.shippingandfreightresource.com/pallet-wide-containers/ (accessed: 10th February 2021).

Sitko, A., ed. 1970. *Jednostki paletowe w obrocie towarowym*. Warszawa: Wydawnictwa Komunikacji i Łączności.

Sitko, A., ed. 1974. *Kontenerowy system transportowy*. Warszawa: Wydawnictwa Komunikacji i Łączności.

SpedCont. 2014. www.spedcont.com.pl/kont_abc2.htm (accessed: 26th July 2014).

Statista. 2021a. *Capacity of Container Ships in Seaborne Trade from 1980 to 2020*. www.statista.com/statistics/267603/capacity-of-container-ships-in-the-global-seaborne-trade (accessed: 10th February 2021).

Statista. 2021b. *Containerumschlag Index Weltweit*. https://de.statista.com/statistik/daten/studie/219285/umfrage/containerumschlag-index-weltweit/ (accessed: 22nd March 2021).

Statista. 2021c. *Major Global Marine Terminal Operators*. www.statista.com/statistics/325934/major-global-marine-terminal-operators/ (accessed: 10th February 2021).

Statista. 2021d. *Number of Ship Losses Between 2010–2019*. www.statista.com/statistics/236250/looses-of-ships-worldwide (accessed: 20th March 2021).

Statista. 2021e. *Number of Ships of APM-Maersk in December-2011*. www.statista.com/statistics/199366/number-of-ships-of-apm-maersk-in-december-2011 (accessed: 6th March 2021).

Statista. 2021f. *Statistiken zum Thema Containerschifffahrt, Containerschifffahrt 2020*. https://de.statista.com/themen/4310/containerschifffahrt/ (accessed: 18th February 2021).

Statista. 2021g. *World Seaborne Trade—Carried by Containers 2017*. www.statista.com/statistics/253987/international-seaborne-trade-carried-by-containers/ (accessed: 10th February 2021).

Statista. 2020h. www.statista.com/statistics/325943/throughput-of-psa-international/ (accessed: 26th April 2021).

Stokłosa, J., T. Cisowski, and A. Erd. 2014. Terminale przeładunkowe jako elementy infrastruktury sprzyjające rozwojowi łańcuchów transportu intermodalnego. *Logistyka* No. 3.

Sweeney, J. 1999. The Prism of COGSA. *Journal of Maritime Law and Commerce* Vol. 30, No. 4.

Synaptic. n.d. https://synaptic.pl/system-tos-obsluga-kontenerow-i-terminala-przeladunkowego (accessed: 28th March 2021).

Szczepaniak, T., ed. 1996. *Transport międzynarodowy*. Warszawa: Państwowe Wydawnictwo Ekonomiczne.

Szumański, A., ed. 2007. *Prawo papierów wartościowych* vol. 18. Warszawa: C.H.Beck.

Szwankowski, S. 2000. *Funkcjonowanie i rozwój portów morskich*. Gdańsk: Wydawnictwo Uniwersytetu Gdańskiego.

Szyszko, M. 2021. *Historia kontenerowych przewozów morskich na świecie*. http://smp.am.szczecin.pl/Content/1028/M.%20Szyszko%20Historia%20kontenerowych%20przewoz%C3%B3w%20na%20%C5%9Bwiecie.pdf (accessed: 4th February 2021).

Textainer. 2021. www.textainer.com (accessed: 12th July 2021).

Tideworks. 2019. *Spinnaker*. www.tideworks.com/products/spinnaker/ (accessed: 24th May 2022).

Tideworks. 2021a. *State of Port Automation*. https://info.tideworks.com/state-port-automation (accessed: 10th April 2021).

Tideworks. 2021b. *Terminal View*. https://tideworks.com/terminal-view/ (accessed: 12th April 2021).

TIL Terminals. 2021. www.tilgroup.com/terminals (accessed: 16th February 2021).

TIR Convention. 2019. https://tir.zmpd.pl/aktualnosci_pliki/f-TIR-445-836-3740.Konwencja_TIR_stan_na_03.02.2019_calosc.pdf (accessed: 22nd March 2021).

Tradecorp. 2021. *Transiflats*. https://tradecorpshippingcontainers.hk/blog/transiflats/ (accessed: 10th February 2021).

Trans.info. 2021. *Container Transport—What Kind of Vehicles Can Be Used by the Carriers [REVIEW]*. https://trans.info/pl/container-transport-what-kind-of-vehicles-can-be-used-by-the-carriers-review-167752 (accessed: 10th February 2021).

Triton. 2020. *Annual report 2019*. Bermuda.

Triton. 2021. www.tritoninternational.com (accessed: 12th July 2021).

Tubielewicz, A., and M. Forkiewicz. 2013. Porty morskie jako element infrastruktury krytycznej łańcucha dostaw. *Logistyka* No. 2.

Tubielewicz, A., and R. Miler. 2014. Determinanty rozwoju logistyki morskiej oparte na zmieniającej się roli portów morskich. *Prace Naukowe Wyższej Szkoły Bankowej w Gdańsku* Vol. 35.

UN Conference. 1980. *United Nations Conference on a Convention on International Multimodal Transport*, United Nations Conference on Trade and Development, held at Geneva from 12th to 30th November 1979 (first part of the session) and from 8th to 24th May 1980 (resumed session).

UNCTAD. 1985. *Review of Maritime Transport 1985*. Geneva.

UNCTAD. 1990. *Review of Maritime Transport 1990*. Geneva.

UNCTAD. 1995. *Review of Maritime Transport 1995*. Geneva.

UNCTAD. 2001. *Review of Maritime Transport 2001*. Geneva.

UNCTAD. 2003. *Review of Maritime Transport 2003*. Geneva.

UNCTAD. 2006. *Review of Maritime Transport 2006*. Geneva.

UNCTAD. 2019. *Review of Maritime Transport 2019*. Geneva.

UNCTAD. 2020. *Review of Maritime Transport 2020*. Geneva.

UNCTAD/SDTE/TLB/227. 2001. *United Nations Conference on Trade and Development Implementation of Multimodal Transport Rules*, Report prepared by the UNCTAD Secretariat, UNCTAD/SDTE/TLB/227, June 2001.

UNCTAD Stat. 2020a. *Container Port Throughput*, annual. https://unctadstat.unctad.org/wds/TableViewer/tableView.aspx?ReportId=13321 (accessed: 10th February 2021).

UNCTAD Stat. 2020b. *Liner Shipping Connectivity Index*, quarterly. https://unctadstat.unctad.org/wds/TableViewer/tableView.aspx?ReportId=92 (accessed: 10th March 2021).

UNCTAD Stat. 2020c. *World Seaborne Trade by Types of Cargo and by Group of Economies*, annual. https://unctadstat.unctad.org/wds/TableViewer/tableView.aspx?ReportId=32363 (accessed: 10th February 2021).

United Nations. 1969. *Problems of Human Environment—Report of the Secretary—General*. https://digitallibrary.un.org/record/729455 (accessed: 24th February 2021).

United Nations. 2012. *The Future We Want. Outcome Document of the United Nations Conference on Sustainable Development*. https://sustainabledevelopment.un.org/content/documents/733FutureWeWant.pdf (accessed: 2nd March 2022).

United Nations. 2015. *Transforming Our World: The 2030 Agenda for Sustainable Development*. www.un.org/en/development/desa/population/migration/generalassembly/docs/globalcompact/A_RES_70_1_E.pdf (accessed: 2nd March 2022).

United Nations. 2020. *World Economic Situation and Prospects as of Mid-2020*. New York: Department of Economic and Social Affairs.

United Nations Economic Commission for Europe. 2021. *IMO/ILO/UNECE Code of Practice for Packing of Cargo Transport Units* (CTU Code). https://unece.org/transportintermodal-transport/imoilounece-code-practice-packing-cargo-transport-units-ctu-code (accessed: 10th February 2021).

United Nations Economic Commission for Europe. n.d. *European Agreement Concerning the International Carriage of Dangerous Goods by Inland Waterways* (ADN). http://staging2.unece.org.net4all.ch/fileadmin/DAM/trans/danger/publi/adn/agreement_text.pdf (accessed: 10th February 2021).

Urbanyi-Popiołek, I., ed. 2012. *Ekonomiczne i organizacyjne aspekty transportu morskiego*. Bydgoszcz: Wydawnictwo Uczelniane Wyższej Szkoły Gospodarki w Bydgoszczy.

Valdes, R.C. 2007. *Los 50 años del inviento del contenedor*. https://revistamarina.cl/revistas/2007/1/claro.pdf (accessed: 3rd February 2021).

Vanroye, K., and B. van Mol. 2008. Parlament Europejski, Dyrekcja Generalna ds. Polityki Wewnętrznej Unii Europejskiej, *Zmieniająca się rola portów morskich UE w globalnej logistyce morskiej: możliwości, wyzwania i strategie*, Parlament Europejski, 2010, https://data.europa.eu/doi/10.2861/86396 (accessed: 17th April 2021).

Wagner, N. 2014. Zastosowanie indeksów frachtowych jako mierników aktywności gospodarczej. *Zeszyty Naukowe Uniwersytetu Szczecińskiego. Problemy Transportu i Logistyki* No. 26.

Waldmann, M. 2016. Rola depotów kontenerowych w obrocie intermodalnymi jednostkami transportowymi. *Autobusy* No. 12.

Waldmann, M. 2017. Terminal kontenerowy jako uczestnik wymiany komunikatów w procesie transportu intermodalnego. *Autobusy* No. 6.

WER-SAD. 2021. *Czym jest agencja celna? Dostępne usługi i koszty obsługi celnej*. http://wer-sad.pl/agencja-celna-uslugi-i-ceny/ (accessed: 22nd March 2021).

White Paper on the Future Development of the Common Transport Policy. 1993. European Commision Publication. https://op.europa.eu/en/publication-detail/-/publication/67d2cd43-9740-42b0-8ba8-e759d36f3109 (accessed: 17th March 2022). Wikipedia.org. 2019. The List of the Largest Container Vessels in the World. https://pl.wikipedia.org/wiki/Lista_najwi%C4%99kszych_kontenerowc%C3%B3w (accessed: 12th February 2021).

Wiśnicki, B., ed. 2006. *Vademecum konteneryzacji. Formowanie kontenerowej jednostki ładunkowej*. Szczecin: Link I Maciej Wędziński.

Wiśnicki, B. 2010. Uwarunkowania obsługi kontenerów 45-stopowych w portach europejskich. In *Polska gospodarka morska—Restrukturyzacja. Konkurencyjność. Funkcjonowanie. Rozwój*, ed. H. Salmonowicz. Szczecin: Wydawnictwo Kreos.

Wiśnicki, B., and I. Kotowska. 2009. Współczesny nadzór techniczny nad kontenerową jednostką ładunkową. *Logistyka* No. 1.

Wojewódzka-Król, K., and E. Załoga, eds. 2022. *Transport. Tendencje zmian*. Warszawa: WN PWN.

WPCI. 2009. World Ports Climate Initiative, *Carbon Footprinting for Ports—Guidance Document*. www.wpci.nl (accessed: 22nd March 2022).

WRI/WBCSD. 2006. World Resources Institute and World Business Council for Sustainable Development, *The Greenhouse Gas Protocol: A Corporate Accounting and Reporting Standard, Revised Edition*. www.wri.org (accessed: 22nd March 2022).

Wronka, J. 2013. Nowy etap rozwoju transportu kombinowanego w Polsce? In *Transport intermodalny w Polsce uwarunkowania i perspektywy rozwoju*. Uniwersytet Szczeciński, *Zeszyty Naukowe no. 778, Problemy Transportu i Logistyki* No. 22.

Wydro, K. 2005. Telematyka—znaczenie i definicje terminu. *Telekomunikacja i Techniki Informacyjne* No. 1–2.

X Change. 2021. *Container Depot Processes explained.* https://container-xchange.com/blog/container-depot/(accessed: 17th April 2021).

Yahalom, S. Z., and Ch. Guan. 2021. *Containership Bay Time and Gantry Crane Productivity: Are They On the Path of Convergence?* http://onlinepubs.trb.org/onlinepubs/conferences/2018/CMTS/MTS_PDF/13A.ShmuelYahalom.pdf (accessed: 22nd March 2021).

Załoga, E. 2014. W kierunku nowego paradygmatu rozwoju transportu w Unii Europejskiej. *Logistyka* No. 2.

Zatouroff, J., and J. Luke. 2013. *Global Shipping.* London: KPMG LLP.

Zhang, R., Z. Jin, Y. Ma, and W. Luan. 2015. Optimization for Two-Stage Double-Cycle Operations in Container Terminals. *Computers and Industrial Engineering* Vol. 83, No. C, May 2015, published online: www.sciencedirect.com/science/article/abs/pii/S0360835215000649 (accessed: 22nd March 2021).

Index